Space - Time and Beyond II

The Series: Episode 2 "Dark Energy"

By

Jack Sarfatti

ISBN: 1-4033-9021-5(E-book)
ISBN: 1-4033-9022-3(Paperback)
ISBN: 1-4033-9023-1(Rocketbook)

This book is printed on acid free paper.

1st Books - rev. 11/14/02

Space-Time and Beyond II

The Series: Episode 2, Dark Energy

By Jack Sarfatti

Jack Sarfatti and Suky Sedgwick (at piano) [1]

Collage 2000 by Allen Cohen, Editor San Francisco Oracle, Haight-Ashbury, 1968

---

[1] "Edie", by Jean Stein, edited by George Plimpton of the Paris Review, Alfred Knopf, 1982. Kyra Sedgwick, married to Kevin Bacon, is Suky's cousin and greatly resembles her.

Table of Contents

From Herbert Gold's "Bohemia"

"The Bohemian physicist ... contributes a balanced scientific non establishment for this expanding society. I don't mean to disparage the work; either...among all the blatherers there sometimes appears a breakthrough thinker. Originality has always required a fertile expanse of fumble and mistake. That's the beauty of the option. Your wastrel life might turn out to be just what's required to save the planet. ... Sarfatti's Cave is the name I'll give to the Caffe Trieste in San Francisco, where Jack Sarfatti, Ph.D. in physics, writes his poetry, evokes his mystical, miracle-working ancestors, and has conducted a several-decade-long seminar on the nature of reality and his own love life to a rapt succession of espresso scholars." *Bohemia, Where Art, Angst, Love and Strong Coffee Meet*, Herbert Gold, p. 14 (Simon and Schuster, 1993)

Herbert Gold[2] and Jack Sarfatti 2000

Chi son?
Sono un poeta.
Che cosa faccio? Scrivo.
E come vivo? Vivo!
In povertà mia lieta
scialo da gran signore
rime ed inni d'amore.
Per sogni e per chimere
e per castelli in aria,
l'anima ho milionaria.[3]

---

[2] Herbert Gold was Jack Sarfatti's literature professor at Cornell in the late 1950's. Gold replaced Vladimir Nabokov who was writing the film script for *Lolita* with James Mason. Mason was a friend of Princess T's father who acted on the stage in Gilbert and Sullivan in London where they met as young men.
[3] "Who am I? I am a poet. What do I do here? I Write. And how do I live? I live in my contented poverty, as if a grand lord, I squander odes and hymns of love. In my dreams and reveries, I build castles in the air, where in spirit I am a millionaire." Rudolfo, La Boheme, Puccini

Jack Sarfatti (Alfred) and Tim Jerome (Frank)[4], Fledermaus, 1964 Highfield - Oberlin Summer Musical Theater, Cape Cod

---

[4] http://www.broadwayusa.org The cast included David Green and Kirsten Falke. See also Cornell Savoyards, Princess Ida, 1958 http://www.rso.cornell.edu/savoyards/58prin.htm , Jack Sarfatti as Prince Hilarion and Alice Bernstein (Jeremy's sister) as Princess Ida. Cast included Ronnie Peierls, son of Sir Rudolph Peierls.

Viva Caffe Trieste by Suky Sedgwick, 1984

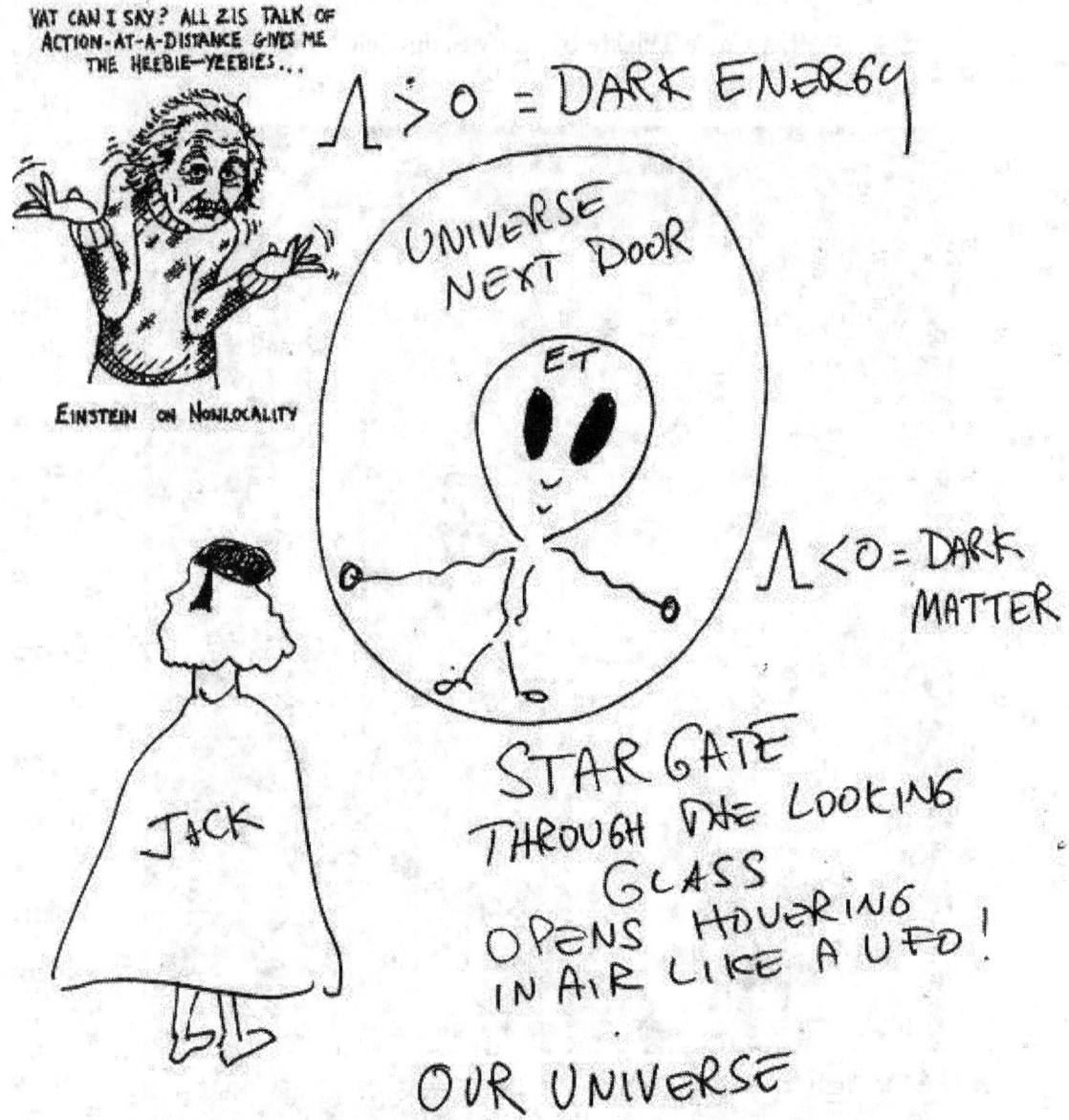

"Faint light appears in the air a few feet above a dirt road; light grows in intensity becoming very bright; bright light then becomes a hole that opens up (growing from 1 to 3 feet diameter) and from within which another light is emanating; a large, black creature (~ 400 lbs., 8 to 9 feet tall) is seen crawling out of the hole (as seen through 3rd generation military night vision, hole appeared 3-dimensional with tunnel-like interior), it stood up and ran away into the surrounding dark of night; the brightly lit hole closed and faded away"[5] Dr. Eric Davis, NIDS report http://198.63.56.18/pdf/davis_mufon2001.pdf

---

[5] See Star Trek IV "The Voyage Home" new DVD release, Spring 2003.

ET From The Universe Next Door by Jack Sarfatti, 2002

By James C. Anderson

"MIDWAY upon the journey of our life

I found myself within a forest dark,
For the straightforward pathway had been lost"

"Nel mezzo del cammin di nostra vita
mi ritrovai per una selva oscura
ché la diritta via era smarrita"

"All else will I relate discover'd there."
Dante, The Divine Comedy

To Princess T Beatrice Sofia and Andrew With Love.

Rodolfo[6]
...o dolce viso
di mite circonfuso
alba lunar
in te, vivo ravviso
il sogno ch'io vorrei
sempre sognar!

---

[6] http://bohemian-opera.tripod.com/boheme_libretto_act1.html "Oh, how sweet your face looks,
its beauty softly kissed by the gentle moonlight. In you, sweet maiden, I see the dreams of love I have
dreamt about forever" http://www.smh.com.au/articles/2002/07/14/1026185139194.html Audio clips of Pavarotti as
Rudolfo http://music.barnesandnoble.com/search/product.asp?ean=28942104921
http://www.cduniverse.com/productinfo.asp?style=music&PID=1023427&frm=sh_google

To Carlo Suares
Who saw them,
"The Woman and The Child"
Coming back from the future
In 1973 in Paris.

By Jack Sarfatti[7]

To Roger Coolidge whose vision and high intelligence made this book possible.[8]

---

[7] The "Two Way Relation" of Action-Reaction (Back-Action) of Bohm and Hiley's "The Undivided Universe" p. 30 & Ch XIV (Routledge, 1993) If the archetype fits, claim it. See the Old Testament stories of Jacob and The Coat of Many Colors and also Joseph and His Brothers as well as Moses.
[8] Cover art by James C. Anderson.

La Boheme Caffe Trieste Cartoons

By Norman Quebedeau
Who also made the FLASH Animations

LIFE AMONG THE CAPPUCCINO PEOPLE

Jack with right hand raised at telephone.

Flash Animations of Star Man Jack and Princess T

http://stardrive.org/cartoon/MagicBean.html
http://stardrive.org/cartoon/spectra.html
http://stardrive.org/cartoon/USKron.html
http://stardrive.org/cartoon/bovines.html
http://stardrive.org/cartoon/dan.html
http://stardrive.org/cartoon/coffee.html
http://stardrive.org/cartoon/Saturn.html

Collage by North Beach Citizen Louis Dinardi

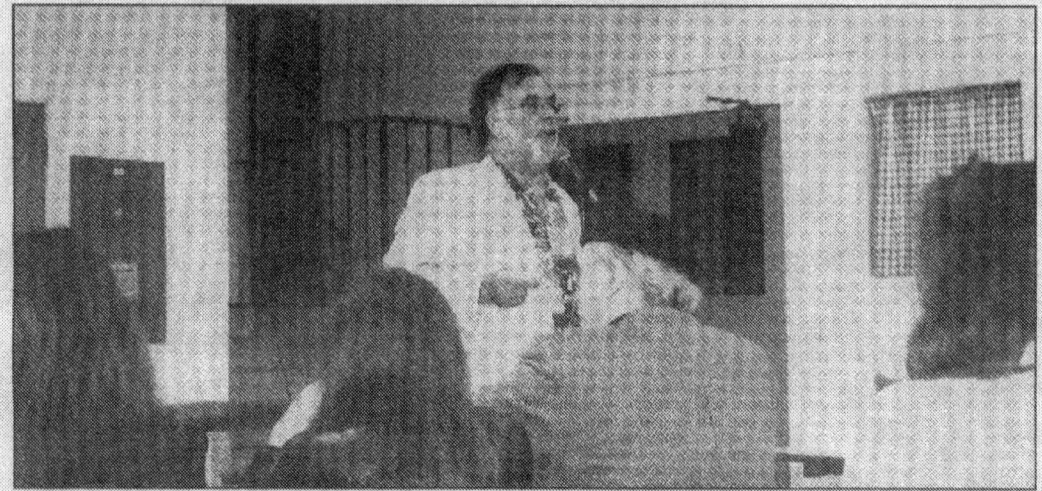

NORTH BEACH JOU

VOLUME 3 NUMBER 1   SEPTEMBER 2002

"On the bottom of the pyramid there are people thrown into homelessness."

Francis Ford Coppola heard support and criticism at a meeting about his North Beach Citizens initiative.

Coppola's Citizens Critiqued

Francis Ford Coppola[9] and Jack Sarfatti 2002

Destiny by Gregory Corso
They deliver the edicts of God
without delay
And are exempt from apprehension
from detention
And with their God-given
Petasus, Caduceus, and Talaria
ferry like bolts of lightning
unhindered between the tribunals
of Space and Time

---

[9] Francis Ford Coppola's charity to support the homeless of North Beach in San Francisco. Dinardi made an award winning film noir 20 years ago with me and many other North Beach characters. The film was shot by Kim Burrafato with Louis directing. The film disappeared in Columbia taken by one of the San Francisco Art Institute students who worked on the production. I was a Fellow of Roy Ascott's Center for Critical Inquiry at the time at SFAI with Angela Davis, Hazel Henderson and Philip Lamantia. Lamantia gave me the copy of "Rashi" about my alleged ancestor of 1000 years ago Rashi de Troyes (AKA Solomon ha-Zarfati 1040 - 1105AD) http://www.charm.net/~brooklyn/People/PhilipLamantia.html

The Messenger-Spirit
in human flesh
is assigned a dependable,
self-reliant, versatile,
thoroughly poet existence
upon its sojourn in life

It does not knock
or ring the bell
or telephone
When the Messenger-Spirit
comes to your door
though locked
It'll enter like an electric midwife
and deliver the message

There is no tell
throughout the ages
that a Messenger-Spirit
ever stumbled into darkness

Down and Out North Beach Citizens Jack Sarfatti and Gregory Corso

On Catholic Church Steps across street from Caffe Trieste with Gisela Mann in the early 1980s between shoots of Louis Dinardi's Film Noir

The Messenger-Spirit requires macro-quantum signal nonlocality in order to exist.

S.M.I².L.E.[10]

Based on the acronym coined by Dr. Timothy Leary[11],
S.M.I2.L.E. is a feature film which aims to discuss
some of the major hurdles facing our species' survival,
along with potential solutions.

We hope to infuse potent information
within an accessible, delicious, sensoryscape.

Space **M**igration
Intelligence **I**ncrease
Life **E**xtension

Topics to be covered include:

Politics of Consciousness

Science and Technology
(ethical application of these tools)

Media and the Universal Mind

Completed Interviews:

---

[10] Jack, this film producer (attached) is trying to reach you. He is traveling, so best way to reach him would be bdub72@hotmail.com. - Amara 10/3/02

[11] Curiously enough in an act of precognitive remote viewing, Tim Leary predicted my discoveries in his book "Exo-Psychology" agented by former DIA spook, The Late Great George Koopman. See Saul-Paul Sirag's article in "Destiny Matrix". Oddly the late Heinz Pagels, a close friend of Nick Herbert and Ira Einhorn died falling off a mountain exactly how he describes at the end of his book, "The Cosmic Code". Pagels forcefully debunked precognitive remote viewing and my advocacy of it. The irony is obvious.
http://www.disinfo.com/pages/article/id773/pg1/ http://www.deoxy.org/8circuit.htm
http://www.caus.org/guestcomments/gc100300.shtml http://www.heart7.net/mcf/mindnet/mn202a.htm
See also Ken Kesey http://www.key-z.com/books.html "Spit in the Ocean" and "Communication With Higher Intelligence."

Paolo Soleri
Sasha, Ann Shulgin
Alex Grey
Erik Davis
Roger Steffens
Curt Joy
Zoe 7, Silvia Polivoy
Morgan Brent
Jon Hanna
Charles Hayes

Fruitfully searching for time with:

Ralph Nader
Jello Biafra
Noam Chomsky
Ray Kurzweil
Jaron Lanier
Michio Kaku
Laura Huxley
Robert Venosa, Martina Hoffmann
Jack Sarfatti
Kat Mckenna

We will commence editing the piece in mid-October,
and are looking to begin distribution in early April.
Thanks in advance,
from the S.M.I2.L.E. crew!

What's a Shaman?

BACK FROM SPACE-TIME AND BEYOND

FRED ALAN WOLF
JACK SARFATTI

ILLUSTRATED BY
JAMES C. ANDERSON

Episode 1: Dancing In The Light[12]

---

[12] Is consciousness the origin of all space-time? Perhaps because space-time emerges from the thought-like cohering of the random quantum zero point fluctuations primarily of the electron-positron field. Are we intimately connected to all the parallel brane worlds of Super Cosmos in M-Theory? Perhaps.

From the original "Space-Time and Beyond", 1975 before I knew of my alleged ancestor, the Cabalist and Torah -Talmudic Scholar, Solomon ha-Zarfati (AKA Rashi de Troyes, 1040 – 1106) who wrote The Commentary that all subsequent Torah -Talmudic Commentaries are based upon.[13]

SCIENTIFIC COMMENTARY

Dr. JACK SARFATTI

I actually coined the title "Space-Time and Beyond". I now see a conference with this exact title at http://www.neugalu.ch by Rene Stettler (stettler@centralnet.ch) sponsored by the Canton and City of Lucerne. The speakers include David Finkelstein and Henry Stapp who I brought down to Esalen in January 1976 as described by Gary Zukav in "The Dancing Wu Li Masters". Also included is Greg Benford. We were graduate students together at UCSD in the mid-60's.

"WITH SUPPORT FROM CANTON AND CITY OF LUCERNE, THE FEDERAL OFFICE OF CULTURE AND SWISS NATIONAL SCIENCE FOUNDATION AND MANY PRIVATE SPONSORS THE
5th Biennial International Symposium of Science, Technics and Aesthetics
WILL TAKE PLACE
ON January 18-19, 2003
lucernetheatre, Lucerne / Switzerland
THE TITLE IS:
SPACE, TIME AND BEYOND
RAUM, ZEIT UND JENSEITS
IT'S a collaboration between the New Gallery of Lucerne and lucernetheatre."
(Oct 8, 2002 e-mail)

---

[13] See "Destiny Matrix" (1st Books, 2002) for details of my 1980 meeting with David Padwa at the AAAS, the rich American with the Land Rover in Ram Dass's "Be Here Now".

ACKNOWLEDGEMENTS

YEARS OF STUDY AND RESEARCH INTO CONSCIOUSNESS TRANSFORMATION BECAME FOCUSED RECENTLY AS I WORKED CLOSELY WITH CARLO SUARÈS (WORLD AUTHORITY ON THE ENERGY CODE OF THE QABALA AND AUTHOR OF THE CIPHER OF GENESIS) AND THEORETICAL PHYSICISTS JACK SARFATTI, Ph.D., AND FRED WOLF, Ph.D.

I CANNOT OVEREMPHASIZE THE DRAMATICALLY ENLIGHTENING IMPORTANCE OF SUARÈS' GUIDANCE IN STRUCTURING THESE THOUGHTS, PARTICULARLY THOSE EXPRESSED IN THE FOREWORD AND THE STRUCTURE OF ENERGY SECTION.

THE MATERIAL PRESENTED IN THIS BOOK OFFERS A NEW/OLD OVERVIEW OF UNIVERSAL ORDER. HOPEFULLY, A SENSE OF HARMONY CAN BE FELT BY THOSE WHOSE STRUCTURES ARE BEING STRIPPED AWAY.

Bob Toben, 1975 in the original "Space-Time and Beyond". E.P. Dutton agented by Ira Einhorn.

by Gary Zukav in the 1979 "Dancing Wu Li Masters"

ACKNOWLEDGMENTS

My gratitude to the following people cannot be adequately expressed. I discovered, in the course of writing this book, that physicists, from graduate students to Nobel Laureates, are a gracious group of people; accessible, helpful, and engaging. This discovery shattered my long-held stereotype of the cold, "objective" scientific personality. For this, above all, I am grateful to the people listed here.

Jack Sarfatti, Ph.D., Director of the Physics/Consciousness Research Group, is the catalyst without whom the following people and I would not have met. Al Chung-liang Huang, the T'ai Chi Master, provided the perfect metaphor of *Wu Li*, inspiration, and the beautiful calligraphy. David Finkelstein, Ph.D., Director of the School of Physics, Georgia Institute of Technology, was my first tutor. These men are the godfathers of this book.

In addition to Sarfatti and Finkelstein, Brian Josephson, Professor of Physics, Cambridge University, and Max Jammer, Professor of Physics, Bar-Ilan University, Ramat-Gan, Israel, read and commented upon the entire manuscript. I am especially indebted to these men (but I do not wish to imply that any one of them, or any other of the individualistic and creative thinkers who helped me with this book, would approve of it, page for page, as it is written, nor that the responsibility for any errors or misinterpretations belongs to anyone but me).

I am also indebted to Henry Stapp, Ph.D., Lawrence Berkeley Laboratory, for reading and commenting upon portions of the manuscript, and Elizabeth Rauscher, Ph.D., founder and sponsor of the Fundamental Physics Group, Lawrence Berkeley Laboratory, for encouraging non-physicists to partake of weekly conferences which normally would attract only physicists. In addition to Stapp and Sarfatti,

CAST OF CHARACTERS   / 17

DAVID FINKELSTEIN
  1958 (one-way membrane hypothesis)

QUASARS
  1962 (discovered)

QUARKS
  1964 (hypothesized)

J. S. BELL
  1964 (Bell's theorem)

DAVID BOHM
  1970 (implicate order)

HENRY STAPP
  1971 (nonlocal connections re: Bell's theorem)

STUART FREEDMAN, JOHN CLAUSER
  1972 (Freedman-Clauser experiment)

TWELVE NEW PARTICLES
  1974–1977 (discovered)

JACK SARFATTI
  1975 (superluminal negentropy transfer hypothesis)

ALAIN ASPECT
  1978 (Aspect experiment—in progress)

The entry "Jack Sarfatti" in the original edition of "The Dancing Wu Li Masters" "Cast of Characters" was *removed* in later editions in a rewriting of history right out of George Orwell's "1984". My original idea of superluminal transfer of negentropy is, of course, implicit in the modern theory of quantum teleportation even though a light limited signal is needed to complete the teleportation at the nonlocal micro-quantum level of sub-quantal equilibrium with signal locality. It's a whole new ball game with sub-quantal non-equilibrium and signal nonlocality[14] at the local macro-quantum level of living matter that explains precognitive remote viewing[15] and, indeed, consciousness, itself! Gary Zukav committed A Great Sin in erasing the above entry from his book that I wrote and edited major parts of with him. Historians take note.[16]

---

[14] http://www.fourmilab.ch/rpkp/valentini.html http://arxiv.org/abs/quant-ph/0203049 http://arxiv.org/abs/quant-ph/0112151 http://arxiv.org/abs/quant-ph/0106098 on the works of Antony Valentini that support my theory from a slightly different point of view than my own. Note: I do *not* agree with Valentini's cosmological ideas.

[15] Jim Schnabel, "Remote Viewers" and the "pre-sponse" mind-brain data of Dean Radin and Dick Bierman.

[16] See the Lawry Chickering letter to Richard De Lauer, Under-Secretary DOD in "Destiny Matrix" (1st Books, 2002). Robert Anton Wilson writes of this in "The Cosmic Trigger" as does Martin Gardner in "Magic and Paraphysics" (MIT Technology Review, 1976) in "Science, Good, Bad and Bogus". Gary Zukav did not

With permission from Lorna McLearie who was living with Gary Zukav during the later parts of his writing of Wu Li Masters. I moved out of Gary's apartment on Telegraph Hill so that Lorna could move in.

> From: LJMCLEARIE@aol.com
> Date: Sun Oct 06, 2002 06:54:39 PM US/Pacific
> To: sarfatti@pacbell.net
> Subject: Re: photograph

> "Dear Jack, I'm glad you found the photograph. Our commonality is our respect for the truth and trusting nature. …… If I had not witnessed your time with Gary firsthand, I couldn't have come forward. Everyday, Gary speaks of compassion and karma. He has become a capitalistic evangelist. He placed money above friendship and loyalty. I'm sorry I wasn't more mature when I was involved. I could have asked questions to lessen the pain for us all.

> Love,
> Lorna"

---

succeed in quantum erasing the facts. See also Jagdish Mann's article in "Destiny Matrix" reproduced below. See also the films "Elmer Gantry" http://www.hollywood.com/movies/detail/movie/168992 and "A Face in the Crowd" http://www.hollywood.com/movies/detail/movie/176975

Lorna McLearie and Gary Zukav late 1970's

From: LJMCLEARIE@aol.com
Date: Thu Oct 03, 2002 01:42:09 PM US/Pacific
To: sarfatti@pacbell.net
Subject: Truth

"I would like to be able to say that my final decision to leave Gary was knowing he hadn't kept his promise to you about the royalties. I was standing in Patsy's[17] kitchen in North Beach when I turned to her and said " Maybe I should try and make it work with Gary" she said " I think you need to know something about Gary and then you will finally leave" She then told me about the royalty promise made to you early on. Also, the fact he

---

[17] Patsy Quinn

kept the secret about having a child but not taking the responsibility until he was well known and financially secure writing about the soul. He watched me be the other parent to my sister's daughter and never said a word about his own child. He demanded I see a therapist to stay in the relationship and when the therapist could see Gary's anger towards me the therapist wanted to see me alone. Gary was so controlling. I cried through the sessions not being able to speak. Gary went into a rage and that was the last time we saw a therapist. Gary said 'You have the therapist wrapped around your finger.' He was *Einhorn angry* and I knew there was no future with someone so angry. I had become depressed living with him. On my last day at the North Beach apt moving my boxes out, our Italian landlady stopped me and said 'Gary is not a nice person and you deserve a nice person.' I didn't say much but smiled and turned to her and said, "Thank you." Jack you can take any part of this. It is your story and unfortunately because of Gary's insecurities we spent little time together. My mother said it this morning: 'In some odd way long ago Jack did you a favor' — meaning it finally made me leave Gary."

The Genius And The Golem
By Jagdish Mann[18]

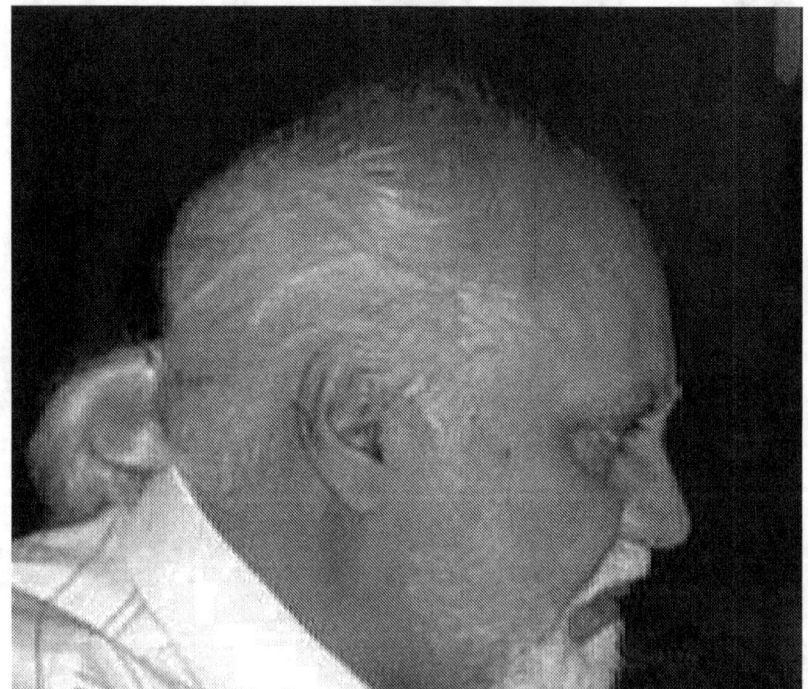

"A man who has seen the unicorn is not to be fooled by a pig."

---

[18] For Bernie Haisch and other Doubting Thomases. "We both met Mike Murphy (the founder of Esalen) who invited us to visit him at Esalen for a few days. As a result we were invited to hold a month-long seminar (in January 1976) on physics and consciousness at Esalen. This was an invitation-only conference in which a dozen or so people participated for various periods of time. ... Gary Zukav was present for a few days ... BTW Jack is right in his claim that he wrote a good deal of this book. Other physicists also assisted including me." Saul-Paul Sirag in email of 10/21/02.

The Dancing Wu Li Masters

I write this is to go on record for what I remember of Gary Zukav and the writing of "The Dancing Wu Li Masters" and the role Dr. Jack Sarfatti played in it.[19] I am in a good place to be a witness to this. I was around from its inception to its finish. In fact, some of the later chapters of the book were written in La Jolla and at my sister's house in El Centro when Csaba Szabo and I had, at Gary's own request, taken him to get him away from North Beach for a month or so.

Major Csaba Szabo, US Special Forces, ret.[20]

---

[19] "Jack Sarfatti, Ph.D., Director of the Physics/Consciousness Research Group, is the catalyst without whom the following people and I would not have met…" Acknowledgments, "The Dancing Wu Li Masters. Gary Zukav reneged on his witnessed oral contract with Jack Sarfatti for 10% of the royalties. Jack's actual role in the writing of the book has been deleted more and more with later editions like the rewriting of history in George Orwell's "1984".

[20] Jack's comment: Gary, like Csaba, did serve in Vietnam and they had that in common. Gary recently cold-shouldered even Csaba – comrade in arms. Part of Gary's disturbance when I met him in 1975 may have been from his military service for which I have some compassion. Gary studied with Henry Kissinger at Harvard and volunteered for Vietnam. Gary said he had been in Black Ops in Cambodia. Although Gary knew no physics at all when I met him, his Harvard education made the difference in his success in addition to the access I provided for him at Esalen and my many hours of working directly with him on the writing of the physics portions of the book including sections from Finkelstein, Stapp et-al. Gary is like a pit bull when he embarks on a project. I met Gary via Tomasso a friend of Francis Ford Coppola's. Francis used to cook us pizzas late at night in Tomasso's wood oven on Kearny off Broadway near Zoetrope and CITY. Tomasso was Gary's roommate and was fed up with Gary. As Csaba said, recalling how Michel Roure threw Gary out of his house in Miami, "To know Gary is to dislike him." Tomasso bribed me with a case of Asti Spumanti to take his room at Gary's. The million-dollar Bay View also helped. So began Gary's meteoric rise to fame and fortune in San Francisco's North Beach.

When Gary first started working on the book, all three of us, Jack, Gary and I lived in North Beach and saw each other almost daily. It was Dr. Sarfatti who tutored him in the intricacies of quantum mechanics. Gary did not know any physics then. His MA was in psychology. There were hours of these tutorial sessions. Many of these took place in the Caffé Trieste and are remembered by many other North Beach residents as well. But I also remember other situations where Gary asked and received technical information from Jack Sarfatti. Gary was nothing if not persistent and Jack nothing if not patient. One incident stands out in my memory Gary was at that time of his life, for all I know, still is, a totally single-minded person and brought an obsessive, and not always attractive, tenacity to his task. On this particular evening, Jack and I were both walking to have dinner with Marcia MacLain who lived on the corner of Green and Kearny when Gary found us. He had come with a purpose and right away barraged Jack with questions. After standing there for ten minutes on the street corner, I reminded Jack of the need to go. Gary just tagged along. As Marcia let us in to her second story flat, he climbed the stairs with us still bombarding Jack with his questions. He carried this on with only a perfunctory acknowledgement of our hostess. Marcia looked at me questioningly and I shrugged, so she put another plate on the table. All through the dinner, Gary stuck to his agenda, speaking between bites. After dinner, I escorted both Jack and Gary to the door and stayed behind to privately apologize to Marcia for having brought an uninvited and unappreciative guest to her carefully planned cooking. The last I saw of them from the second story window was them walking down Kearny Street, engrossed in an animate discussion.

There is another important contribution that Jack Sarfatti made to the writing of "The Dancing Wu Li Masters." He not only tutored Gary enough in physics so that he could ask intelligent questions from other physicists, he even introduced him to people like Dr. Henry Stapp, and took him to Esalen to meet many more.

I cannot say that without Jack Sarfatti, Gary Zukav could not have written "The Dancing Wu Li Masters," but I can say with certainty, that without Jack Sarfatti, it would not have been the same book. It would have been more like Gary's recent work, "The Seat of the Soul," a book full of trite truths and platitudes held together with Talk Show spit.

"You can fool all the people some of the time, and some of the people all the time, but you cannot fool all the people all the time." Abraham Lincoln

The Lost Chord

Shakespeare's Hamlet has the play within the play. Mozart never finished his Requiem. F. Scott Fitzgerald had a heart attack before he could finish "The Last Tycoon". Well I am still here. As an overture I present a fragment of an unfinished book, the sequel to the very silly, but very popular, New Age Airhead Cartoon Book For Dummies, "Space-Time and

Beyond". I wrote it in an altered state of consciousness with Fred Alan Wolf and Bob Toben[21] in Paris under the influence of Cabalist Carlo Suares[22].

Jack Sarfatti and Fred Alan Wolf in Paris 1973

I also worked on it in London during the Uri Geller tests, at Abdus Salam's Physics Institute in Trieste, Italy, in Philadelphia with Ira Einhorn, and in Ossining with Andrija Puharich, in Chicago with Bob Toben and finally in San Francisco in the 1973-74 est Era of "Tales of The City"[23]. Little did we gullible fools know we were The Three Stooges, The Three Amigos, in Ira Einhorn's[24] Psi Wars! Beatnik Physics? — You got it Bhubba! Those were the days my friends we thought they'd never end. Well they did. This is The End of Days, The Lost Weekend. This is The End to Ordinary History. You better believe it. "Apre

---

[21] Equally stoned and drunk on French Vine in our Coming of Age walking through "The Doors of Perception".

[22] Suares's comrades included Krishnamurti, Aldous Huxley and Lawrence Durrell, Liam O'Gallagher http://artscenecal.com/ArtistsFiles/OGallagherL/LOGallagher.html , Beatrice Wood and Robert Rheem http://www.emptymirrorbooks.com/thirdpage/ricebio.html . Suares, a Sephardic Jew from Alexandria, was the inspiration for "Balthazar" in Durrell's "The Alexandria Quartet"

[23] Werner Erhard, Armistead Maupin -- the San Francisco of Herb Caen and Mayor George Moscone.

[24] Esalen celebrity, counter-culture super-star, Harvard Fellow, colleague of Jacques Vallee on UFO investigations, Ira Einhorn was the literary agent for the original "Space-Time and Beyond". A fugitive for 20 years, he is now in prison for the alleged murder of his girl, Holly Maddux. Einhorn is now on trial for murder in Philadelphia as this is being written. For more details see my book "Destiny Matrix" (First Books, 2002)

moi Le Deluge!" said an alleged ancestor of mine! ☺ "Sarfatti" means "The Frenchman" in Hebrew alluding to The Great Rashi des Troyes (1040-1105).

Star Man Jack 2002, Ophiuchus R A 17h 39m 55s D 3° 11'

Photo by AT Conway

I guess it was in 1998 or so, having linked up to funds from Roger Coolidge, a man with the vision and intellect of Alfred Lee Loomis[25], and, unlike Gary Zukav[26] whose career I created at Esalen in 1976, I am generous to my industrious creative friends on The Eagle's Quest. I gave my old buddy from the San Diego State Physics Department[27], Fred Alan Wolf[28], a call. "Fred", I said, "it's time we did *Space-Time and Beyond* again." This time let's do it right. I'll "work on a new and original plan, said I to myself said I"[29]. After all, we are older and, presumably, wiser. No more our debauched youth of wild sex with many exciting women in the 60's and in the 70's "Tales of The City". I persuaded Fred, whose heart was not really in it[30] however, by having my Consigliore, Dennis Wishnie, of Bugatto

---

[25] "Tuxedo Park: A Wall Street Tycoon and the Secret Palace of Science That Changed the Course of World War II" by Jennet Conant, 2002

[26] Author of the "Dancing Wu Li Masters" that I wrote large parts of and helped him closely edit. Gary coldheartedly abandoned *all* of his key friends, male and *female*, who literally took him out of the depths of suicidal depression and made his financial success possible. As soon as he got his money he did not wish to know any of us. That is Gary's real character if the truth be told.

[27] Fred, older by five years was an associate professor, I was an assistant professor. We are parodied as a composite character in the opening scenes of "Ghost Busters".

[28] Winner American Book Award 1982 "Taking The Quantum Leap".

[29] "Trial By Jury" Gilbert and Sullivan

[30] Fred opted out after a few months getting a contract for "The New Alchemy" without me. However, he wrote a speculative paper on the quantum physics of Libet's mind-brain experiments that show a precognition

and Wishnie Law Offices in San Francisco's and Francis Ford Coppola's Italian North Beach, cut Fred a check for an initial hefty advance, which he accepted eagerly.[31] Bob Toben was out of the picture by that time. He was not the same person he was almost 30 years ago. Fred and I, of course, never grew up and we are really still a couple of manic six-year olds in sixty-year-old bodies. We were "The Odd Couple" at San Diego State. Today we are "The Sunshine Boys" with me as Walter Mathau and Fred as Jack Lemmon. Anyway, Fred put a notice up at the San Francisco Art Institute for a Bob Toben replacement and that's how we met James C. Anderson, our Harry Potter, who drove his VW Bus down from Sacramento over The Golden Gate to meet Freddy and me.

Jack, Fred lower right, Saul-Paul Sirag back middle and Nick Herbert 1975

In Francis Ford Coppola's *CITY* "Faster Than The Speeding Photon" by Rasa Gustaitus

---

effect also seen by Dean Radin and Dick Bierman and discussed by Roger Penrose in "The Emperor's New Mind", so I figured he did something useful with the money I advanced him.

[31] I eventually gave Fred about additional funds including travel expenses to the Vigier Conference in Toronto and Stuart Hameroff's consciousness conference Tucson II.

So I loaded up my trusty van with gas and questions
and set off for the fabulous city by the bay
where these two very scientific wizards were perhaps even
now conducting arcane explorations into the unknown.
Who better to pose my eternal questions to?

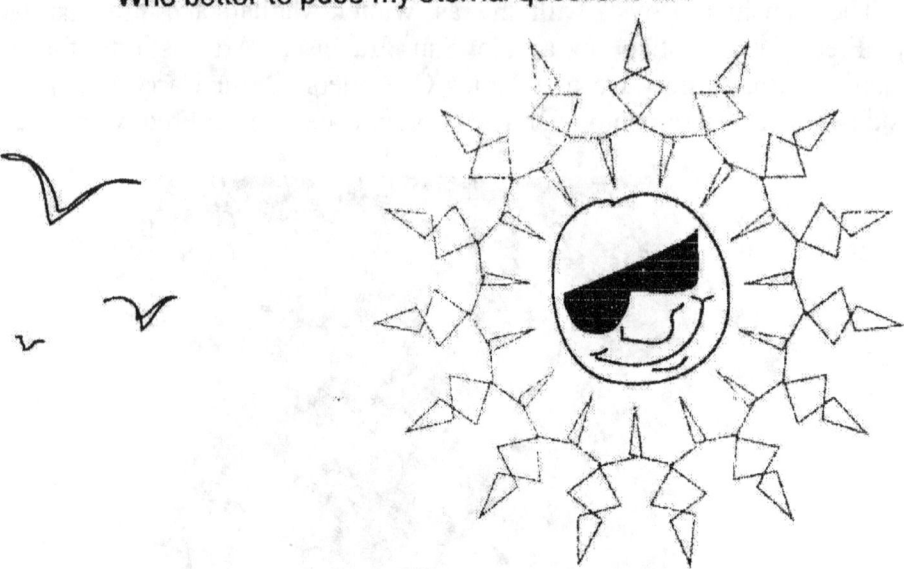

They is they is a couple of wizzes
if ever some wizzes there wazzes...
if i only had a brain
det de det de de de de dah...

When I arrived at their semi-secret sanctuary I admit I was
a bit worried-- everything was dreamlike; I stood
faceing a long dark hallway... strange pulsing light
eminated from the chamber at the far end...
there was the faint sound of etherial voices
speaking in an otherworldly tongue...
What would these mystical quantum wizards
of time and space be like???

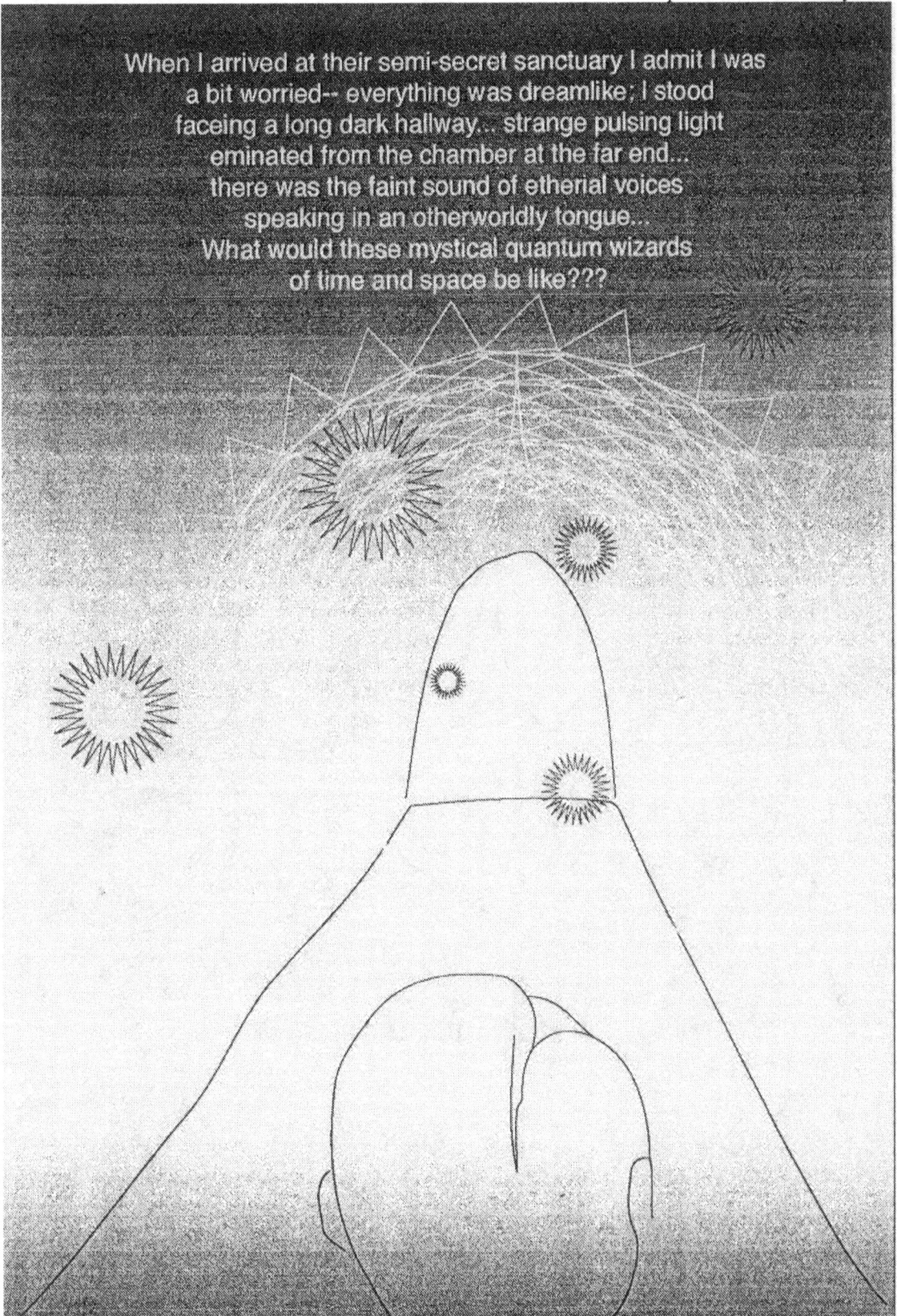

Look, a beamish boy!

Marvelous-- shall we send him away?

Oh, let's best not. It's rather brillig out.

Then shall we have him stay for supper?

Oh let's do. Put him in the pot with the slithy toves.

NOT! KIDDING!!!

I gave James a substantial check for his work and this, based on what I told him I wanted[32], is what he came up with. I wrote the comments under the cartoons. I include some of the original work by Fred Alan Wolf that he did for this project.[33]

---

[32] I have deleted several, *but not all*, of the cartoons based on Fred's quantum ideas, which come from Niels Bohr. I consider Niels Bohr to be the second great Fairy Tale Maker of Copenhagen after Hans Christian Anderson. Bohr's Tale of Puff, The Smoky Dragon was even bought by that great Shaman Physicist, John Archibald Wheeler. I am of the Einstein School with David Bohm of Objective Realism. Of course it is a nonlocal *micro-quantum reality*, but because of coherence of the vacuum giving it the "unbearable lightness of being", the *macroscopic quantum world* of Einstein's space-time curved by energy, is mostly local.

Overture: A Pop Art Lecture on Reality by Fred Alan Wolf

---

[33] A Pop Art Lecture on Reality by Fred Alan Wolf included below. Fred wrote me on 9/22/02 "Jack, if you wish, you can publish your book without listing me at all as a co-author. Please consider the contribution I made to the book as a consultation. That way you can keep all of the royalties it earns and you owe me nothing more. There is no need to send me what you have done for further input."

What is quantum reality?

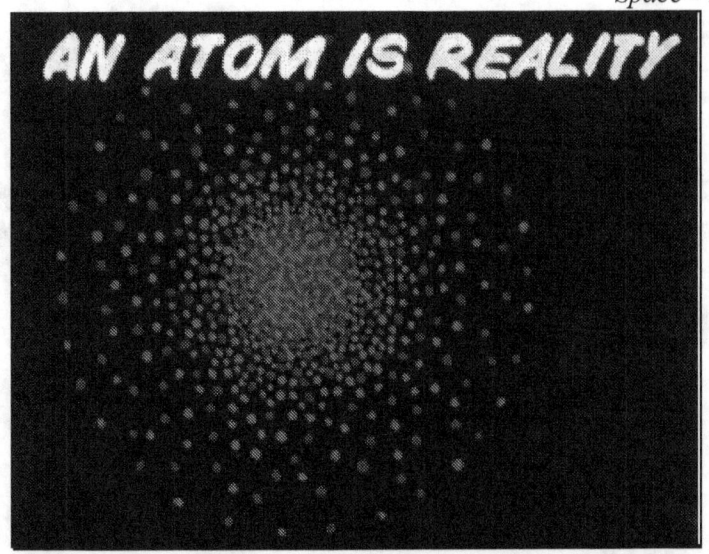

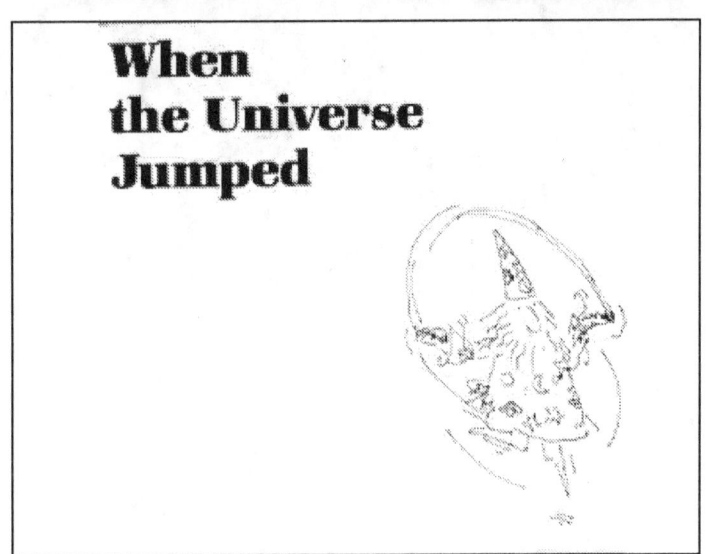

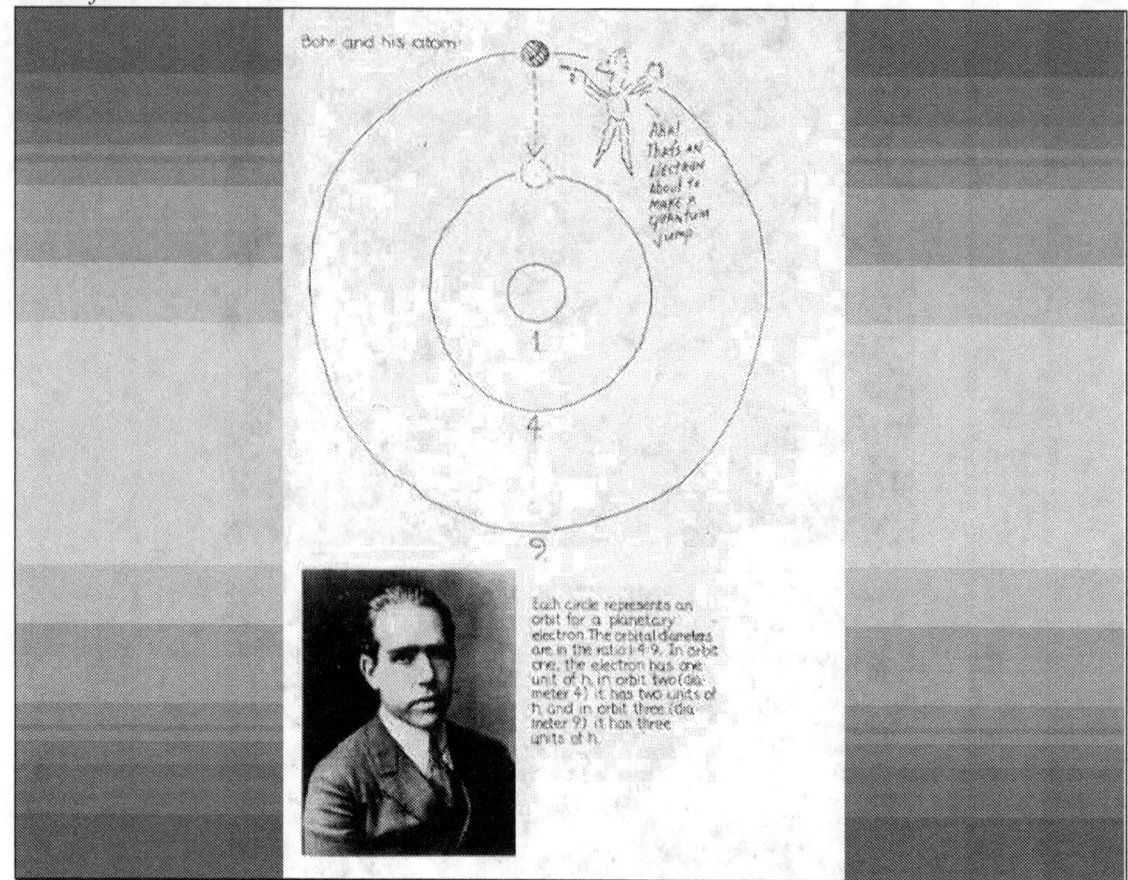

Quantized electron energy levels in Niels Bohr's atom of 1913

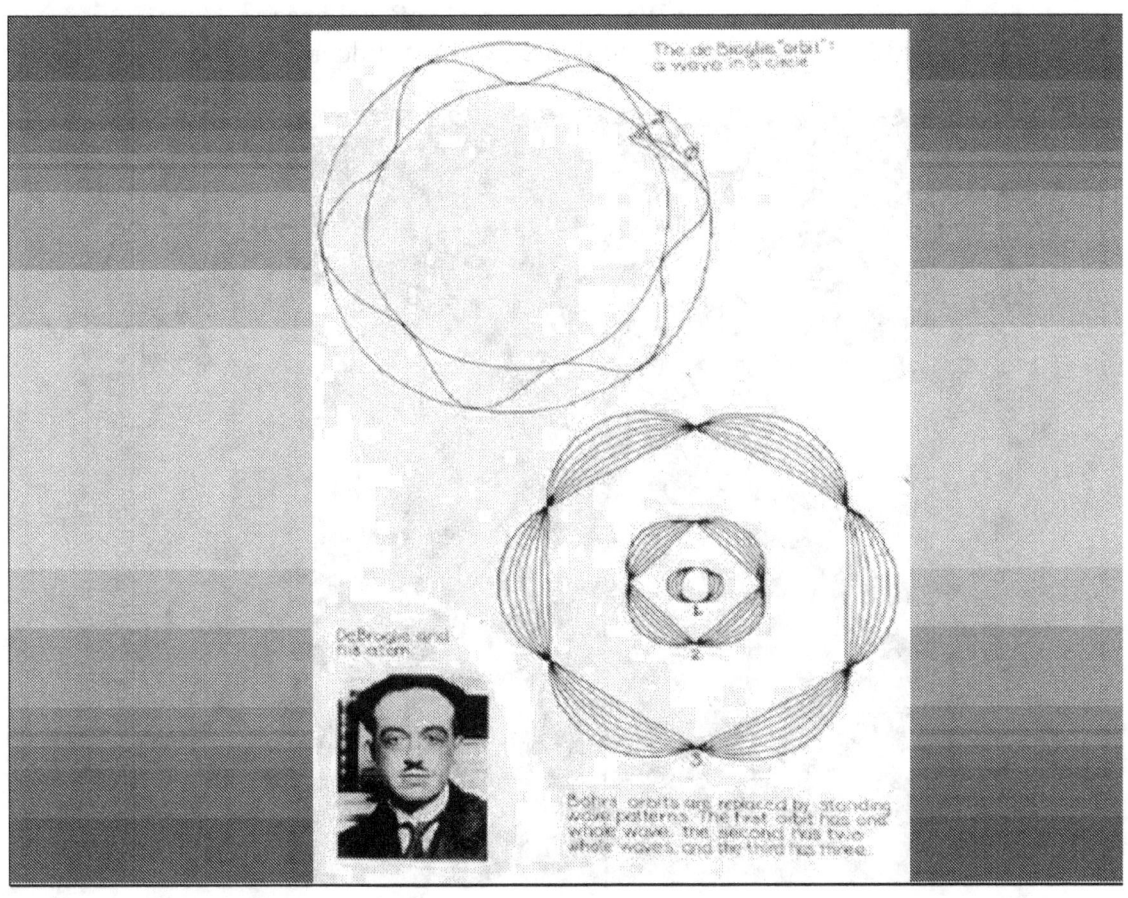

Prince Louis De Broglie's Quantum Matter Waves

Wave or Particle?

Niels Bohr said: Wave with miracle of "collapse".
David Bohm said: Wave and Particle as the "Hidden Variable."

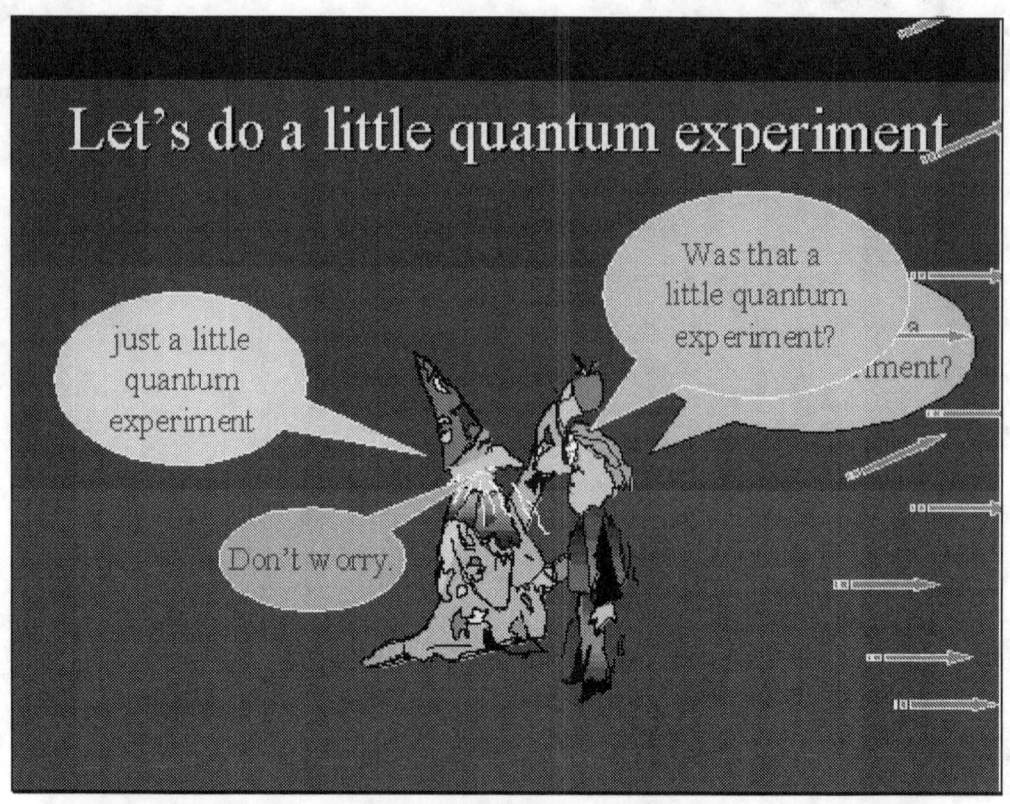

The Incredible Quantum Jumping Cube

Does this cube quantum jump?

End of Fred's Pop Art Lecture[34]

The Meaning of Quantum Theory

By Jack Sarfatti

Since the quantum wave is thought-like, Bohr's position suggests no "there" there. Bohr's "Youinverse" is pure Eastern idealism when taken to its logical conclusion. Einstein, in contrast, always suspected that The Truth is Out There. This was the basis for the Einstein-Bohr Debate still going on today.

What is the meaning of the quantum theory? There is no consensus among the mainstream Physics Pundits. Niels Bohr and Albert Einstein could never agree about this. Richard Feynman said no one understood the meaning of quantum theory. Bohr said anyone who thought he or she understood it, didn't. Taking that risk, going out on a limb, I think I understand the meaning of quantum theory. Basically, it's this. Physical reality divides into both mind and matter. Both are equally real physically. Quantum waves are mind. Elementary particles are matter. Space-time geometry is matter in this sense as is the hyperspace of the parallel brane worlds of Super Cosmos described by Stephen Hawking in his book "The Universe in a Nutshell". Elementary particles like electrons and quarks are what David Bohm called "hidden variables". John Archibald Wheeler, with the characteristic wide-eyed inverted neoteny[35] of genius thought elementary particles were tiny wormholes in the vacuum fabric of space.[36] This can only happen if gravity gets very strong at small distances perhaps as large as a nanometer[37], or even a little larger. M-theory has this possibility. Strong gravity at short range means space-time loses its large-scale stiffness against bending by energy. Think of space-time like a guitar string. At large distances the string is stretched tight giving a high note. At smaller distances the string tension is reduced giving a lower bass note. Evidence for this may be seen in the *alleged* flight of alien extra-terrestrial flying saucers. This is controversial of course. I am not a true believer in flying saucers and alien ETs. I also do not knee-jerk debunk sober reports from qualified observers some in the US Military Intelligence since WWII.

Niels Bohr basically thought that micro-quantum reality was pure mind. There were only micro-quantum waves without Isaac Newton's hard massy tiny particles at the micro-level of less than a nanometer. Micro-quantum waves are different from classical sound waves and classical electromagnetic waves. Micro-quantum waves are not mechanical like classical sound and light. They are not like the separate parts of your automobile engine acting in fixed relationships to each other. Micro-quantum waves are context dependent. They can act faster-than-light "nonlocally" and they exist in higher dimensional space when "entangled"

---

[34] Fred is Goldmund to my Narcissus (Hermann Hesse's novel). Or, he is Dionysus to my Apollo in the sense of Nietzsche's "The Birth of Tragedy".

[35] I.e. staying young and playful with ideas into old age.

[36] This the "geometrodynamics" of "mass without mass" and "charge without charge".

[37] A billionth of a meter. The hydrogen atom in its lowest energy state is about one tenth of a nanometer across. Large Rydberg hydrogen atoms in short-lived excited states are much larger. Technology has advanced. We can now trap even single electrons in "quantum dots", i.e. designer atoms. It may be possible to nano-engineer a conscious artificial intelligence from a "More is different" nano-network of quantum dots with single electrons. Indeed, the living human brain is an example of exactly that inside the hydrophobic cages of protein dimers inside microtubules. See "Shadows of the Mind" by Roger Penrose on Stuart Hameroff's half-baked but sometimes interesting ideas for details.

as in information applications to "teleportation" and "untappable communication command control" (C^3). However, you cannot use micro-quantum waves to communicate a decodable message either faster-than-light or precognitively backwards in time! There is a Catch 22 because micro-quantum waves are uncontrollably random. The random quantum noise means you cannot open the lock hiding the message until you get the key. The key is always a classical signal that cannot move faster-than-light. That is, we have micro-quantum nonlocality with macro-signal locality. There is one other thing about quantum waves, even though they are mental they are not conscious like our minds are!

Slaying The Smoky Dragon

Copenhagen, Denmark has given the world two great fairy tale storytellers, Hans Christian Anderson and Niels Bohr. Bohr's great fairy tale that even seduced John Von Neumann, Eugene Wigner, John Archibald Wheeler and Roger Penrose is that of Puff n' Pop the "Smoky Dragon". This means no hidden variables. In spite of Feynman's diagrams you cannot picture an electron as a tiny something in space moving in a continuous path. John Von Neumann, the man who with Alan Turing, got the key ideas for your PC and Mac, introduced the Deux Ex Machina, the truly supernatural miracle called "Pop" the collapse of the quantum wave into a particle in a definite location in a suitable "measurement". Well that's what a lot of mainstream Physics Pundits believe as a kind of religious faith. Quantum cosmologists don't like this "Copenhagen Interpretation" because you cannot divide the total universe into observer and observed the way you must in Bohr's interpretation. The cosmologists prefer the parallel universe many worlds ideas without Smoky Dragon collapses. Many worlds come in many variations on the thematum.[38] The most recent is David Deutsch's "multiverse" which from my POV means "many minds". The material parallel brane worlds in Hawking's "Nutshell" hyperspace are not the same as the mental multiverse. The Bohmian realist solution is that we have both mental multiverse and materialist parallel brane worlds in hyperspace. My theory of everything is both combined into "Super Cosmos" with the mental multiverse as Hawking's "Mind of God". Hawking wrote:

"Einstein once asked the question: 'How much choice did God have in constructing the universe?' Even if there is only one possible unified theory, it is just a set of rules and equations. What is it that breathes fire into the equations and makes a universe for them to describe? ... Why does the universe go to all the bother of existing? Is the unified theory so compelling that it brings about its own existence? Or does it need a creator, and, if so, does he have any other effect on the universe? And who created him? ... if we do discover a complete theory, it should in time be understandable in broad principle by everyone, not just a few scientists. Then we shall all, philosophers, scientists, and just ordinary people, be able to take part in the discussion of the question of why it is that we and the universe exist. If we find the answer to that, it would be the ultimate triumph of human reason - for then we would know the mind of God." – "A Brief History of Time"

---

[38] Gerald Holton at Harvard "Thematic Origins of Modern Science".

A Visual Poem by Fred Alan Wolf

The anatomy of the spirit

- spirit the vibrations of nothing
- soul   the reflection of spirit at the nodes of time
- matter the reflection of spirit at the nodes of space
- self   the reflection of soul in matter

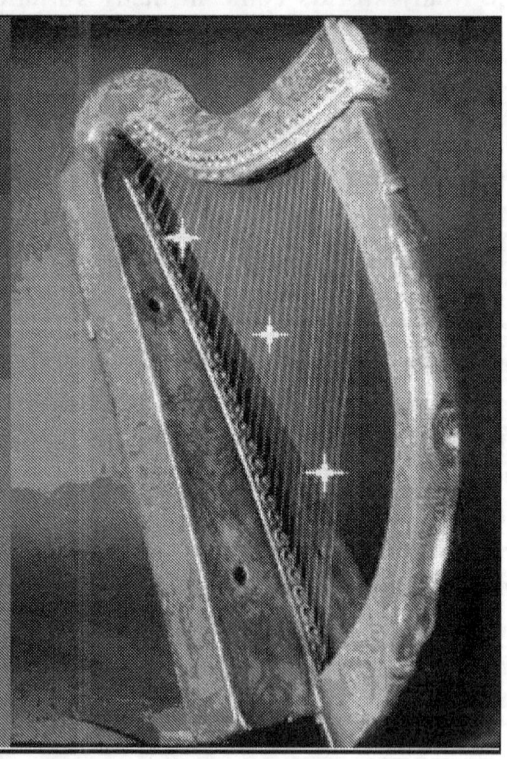

Every-
thing
is
vibration

String Theory

History $\Rightarrow$ NOW $\Leftarrow$ Destiny[39]

---

[39] Jacky's Yeshiva Boy Commentary: This Yakir Aharonov's "Two State Vector" theory of Advanced Destiny meeting Retarded History in the Be Here Now. Rotate 90 degrees to Hawking's imaginary time to get Carlo Suares's Cabala Diagram of the "Two Way Relation" (Bohm and Hiley p. 30) between Aleph and Bayt for the

La Forza del Destino!

Mighty Joe Genius
The Hidden Variable
The Particle
The IT

---

generation of inner consciousness in macro-quantum systems not in sub-quantal equilibrium, hence signal nonlocality violating micro-quantum mechanics. Think of nonlocal incoherent micro-quantum non-mechanics with signal locality as "Old Testament" with local coherent macro-quantum non-mechanics with signal nonlocality as "New Testament". I do not, however, wish to be crucified like Rabbi Yeshua any more than I already have been. ☺

Ψ-Angel

The Mental Quantum Wave
The BIT

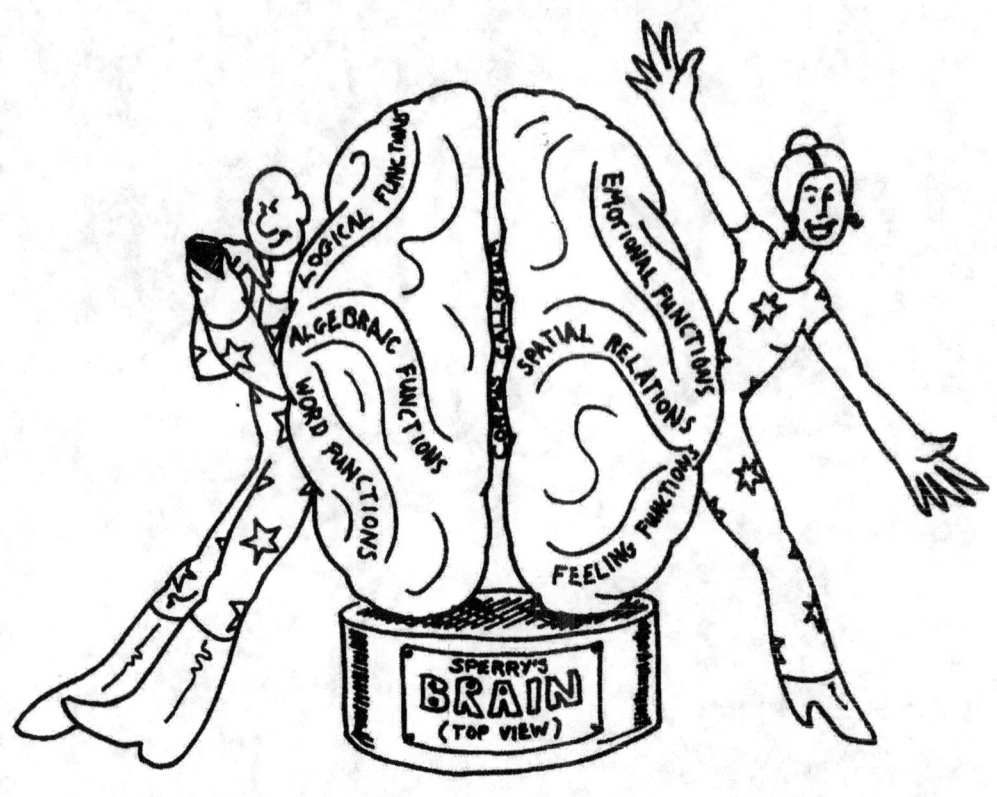

The Macro-Quantum Physics of Consciousness

What about consciousness? What about us? How do we fit in? Wheeler describes Einstein's theory of gravity, in which space-time geometry is curved by energy, as matter getting its marching orders from geometry, which is also in the grip of matter's energy density. Therefore, what we have is a two way street of action and reaction between geometry and matter in Einstein's classical theory of gravity, i.e. the general theory of relativity. David Bohm[40], who worked with Einstein in early 1950's at Princeton, showed an analogy of quantum theory with this two-way action-reaction of general relativity.[41] The mental quantum wave is a "pilot wave". Both classical matter and geometry are "hidden variables" that get their marching orders from quantum waves. Quantum waves are really thought-like "information" waves. Information comes in "bits". There are classical bits

---

[40] I was an Honorary Research Fellow with Bohm at Birkbeck College, University of London in early 1970's. See Martin Gardner's "Magic and Paraphysics" in "Science, Good, Bad and Bogus" for more details on this.

[41] See David Bohm and Basil Hiley's "The Undivided Universe" p. 30 and Ch 14 for more details. Also recent online papers by Antony Valentini for a modern look at this "two way relationship" that permits "signal nonlocality". Micro-quantum signal locality only works when the hidden variables are in thermal equilibrium so that the probability of a micro-quantum event is proportional to the square of the quantum wave. This condition can break down. Indeed, I think it breaks down in all non-equilibrium living matter where we now have "More is different" macro-quantum locality with signal nonlocality.

called "c-bits" and quantum bits called "qubits". In Bohr's theory qubits collapse into c-bits and a quantum computer has massive parallelism from an infinity of parallel c-bit computing processes that combine to solve a problem much faster than any classical computer described by a Turing Machine[42] or any other formal mathematical logical scheme equivalent to it.[43] My old Cornell colleague[44], Stanford's Lenny Susskind, has a "world hologram" model in which volumes of space have a c-bit content equal to the area bounding that volume divided by the Planck area 10^{-66} cm[2]. This reduces all space to the "Bekenstein" c-bits previously thought to only be on the event horizons of black holes that obey the laws of thermodynamics.[45]

David Bohm pointed out that the only reason that micro-quantum waves cannot be used for direct communication, faster than light, slower than light, and even backwards in time[46] is that matter and geometry get their marching orders from mental quantum pilot waves. However, these mental waves do not get any direct feedback or "back-action" from the matter-geometry[47] whose motion they pilot. My consciousness generation conjecture, is that if and when any mental quantum bit pilot wave gets direct feedback from the classical geometrodynamical hidden variable it is piloting, then that mind field acquires inner consciousness with feelings. This seems to require a "More is different"[48] phase transition to a non-equilibrium "macro-quantum" complex system with signal nonlocality that is *not* found in micro-quantum physics.

Bohm and Hiley's key insight

"Finally it should be pointed out that unlike what happens with Maxwell's equations for example, the Schrodinger equation for the quantum field does not have sources, not does it have any other way by which the field could be directly affected by the conditions of the particles. This of course constitutes an important difference between quantum fields and other fields that have thus far been used. As we shall see, however, the quantum theory can be understood completely in terms of the assumption that the quantum field has no sources or other forms of dependence on the particles. We shall in … 14.6, go into what it would mean to have such dependence and we shall see that this would imply that the quantum

---

[42] Alan Turing cracked the Nazi war code at Bletchley Park in WWII that saved Britain from the destruction of Hitler's and Goring's Luftwaffe. My Cornell fellow student from late 1950's, Tom Pynchon writes about this in "Gravity's Rainbow". Turing tragically apparently committed suicide after WWII by eating a poisoned apple in anguish over a developing scandal alleging his homosexuality.

[43] See Stephen Wolfram's "A New Kind of Science" for all the details about this. Thanks to Creon Levit of NASA Ames for explaining Wolfram's approach.

[44] 1963, *joint work* with Susskind and my childhood buddy, Johnny Glogower, on time and phase operators in quantum theory.

[45] See Hawking's "Universe in a Nutshell" for more details.

[46] A much stronger "delayed choice experiment". Wheeler's cannot be used for precognitive remote viewing like the Hal Puthoff- Russell Targ SRI CIA funded experiments in early 1970's with Uri Geller, Ingo Swann and Pat Price. I went there in 1973 to visit in the course of writing the first very silly immature "Space-Time and Beyond" with Bob Toben and Fred Alan Wolf agented by Ira Einhorn.

[47] The material "geometry" hidden variable is 3-dimensional not 4-dimensional. The mental pilot wave of this material geometry lives in an infinite dimensional field configuration space if the 3-dimensional geometry is a continuum.

[48] Coined by P.W. Anderson.

theory is an approximation with a limited domain of validity." P. 30 *The Undivided Universe*, Routledge, 1993

No source for the "thought-like"[49] quantum Ψ field is action of mind on matter without back-action or direct reaction of matter back on mind. Mind grips matter and makes it move randomly unconsciously unintelligently without matter gripping back on mind to awaken it into inner consciousness with creative intelligence apart from hard-wired instinct in the form of naturally selected classical cybernetic feedback-control loops as explained by Norbert Weiner.

Matter receiving its marching orders[50] from mind without matter gripping back on mind is analogous to globally flat special relativity without gravity. In that limited domain matter receives marching orders from space-time geometry without matter bending space-time and geometry into gravity.

No source for the micro-quantum Ψ field[51] also corresponds to Antony Valentini's "sub-quantum heat death" or "sub-quantum equilibrium" in which nonlocal quantum theory has signal locality. Signal locality is the basis of quantum computing, quantum teleportation and untappable quantum cryptography. They all collapse in post-quantum or macro-quantum theory with local emergent order[52], but long range hologram phase coherence, from symmetry breaking in the vacuum for virtual quanta and in the ground state for real quanta. We now have signal nonlocality with local macro-quantum order that corresponds to Valentini's "sub-quantal non-equilibrium". All living matter is non-equilibrium with signal nonlocality violating micro-quantum theory.

Faster than light signals cannot happen in micro-quantum theory with signal locality. Faster than light signals, even backwards in time from the future, do happen as a matter of common fact, in living matter whose signature is signal nonlocality. This is what the "presponse" mind-brain data of Libet, Radin and Bierman is all about. This is what ordinary consciousness is all about. This is what the paranormal is all about. The knee-jerk debunking Pundits who deny this fact cut off their noses to spite their faces. Heinz Pagels was one of those Pundits and he died in precisely the way he precognized at the end of his book "The Cosmic Code" falling off the mountain. This was a tragic way to prove a point.[53]

---

[49] "thought-like" and "rock-like" in sense of wave and particle respectively coined by Henry Stapp.

[50] Image coined by John Archibald Wheeler.

[51] This is the limit of zero back-action with a "fragile Bohm quantum potential".

[52] P.W. Anderson's "More Is Different."

[53] Persi Diaconis might write this off as random chance or suicide. There is always a matter of judgment involved in complex individual cases like this in which common probability reasoning breaks down. Jung and Pauli called this "synchronicity".

Altered States of Consciousness

Our normal waking consciousness
is but one special type of consciousness
whilst all about it parted from it by the filmiest of screens
there lie potential forms of consciousness entirely different.

-- William James

What am I? The Cipher of Genesis

"If you want to know who we are?"

Mikado, Gilbert and Sullivan

"Then he plunged himself into the billowy wave."
Bab Ballads, Sir W.S. Gilbert

The multiplicity of mortal souls
is an illusion...

Individual "selves" are the reflection
of the one eternal soul in matter

Many-One

How Many Souls Are There?

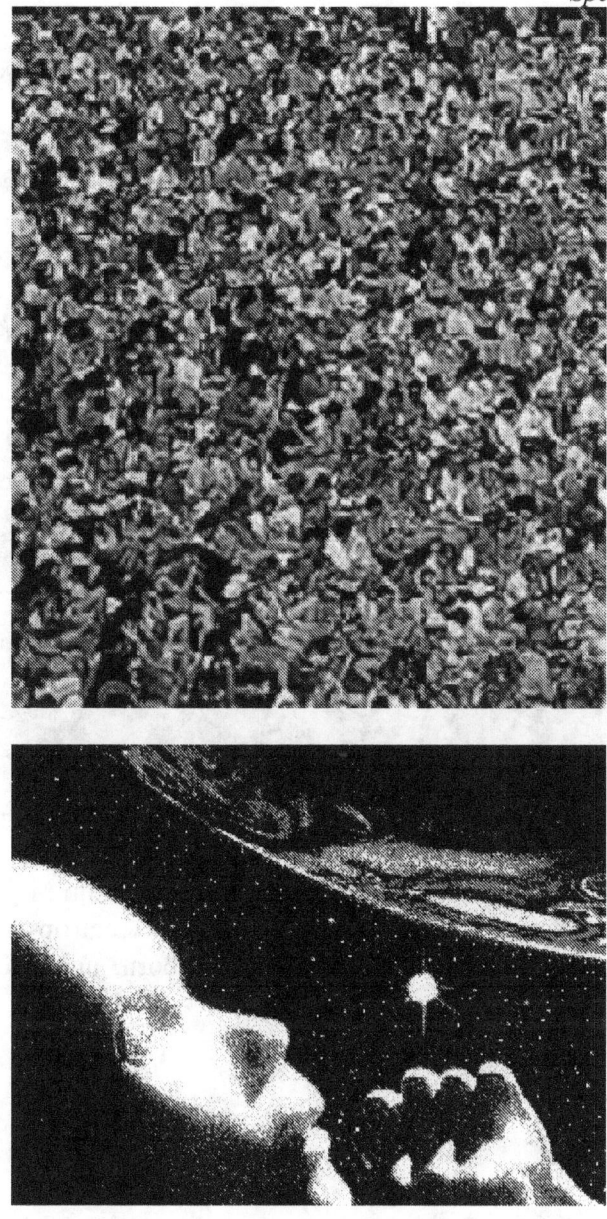

Just One?[54]

The Mind of God and The Cipher of Genesis

The One Eternal Soul is what Stephen Hawking calls "The Mind of God". I actually know what it is. It is the complex number field vacuum coherence local order parameter Ψ that calms the quantum zero point fluctuating turbulence, quieting Heisenberg uncertainty random noise, and in the great silence[55] OM or Aleph of Einstein's cosmological field $\Lambda = 0$, the smooth curved space-time of gravity emerges like Botticelli's *Venus Arising* from the variable macro-quantum wave phase Θ of Ψ.

---

[54] Idea by Fred Alan Wolf.
[55] Descended on the Chaos of Flat Land.

The $\Lambda = 0$ non-gravitating vacuum (e.g., Arthur Miller's "After The Fall" and John Milton's "Paradise Lost") in the early universe post symmetry-breaking[56] vacuum phase transition from unstable higher energy chaotic maximally random high entropy Flat Land to metastable lower energy emergent smooth less random low entropy Curve World explains the thermodynamic Arrow of Time in which we are born, age and die as the universe expands.

---

[56] The spontaneously broken "More is different" (P.W. Anderson's phrase) symmetry is that of the translation subgroup of the Poincare group for Flat Land. This subgroup is generated by the combined energy-momentum 4-vector that John Archibald Wheeler calls "momenergy" in "A Journey into Gravity and Space-Time" (Scientific American, 1990).

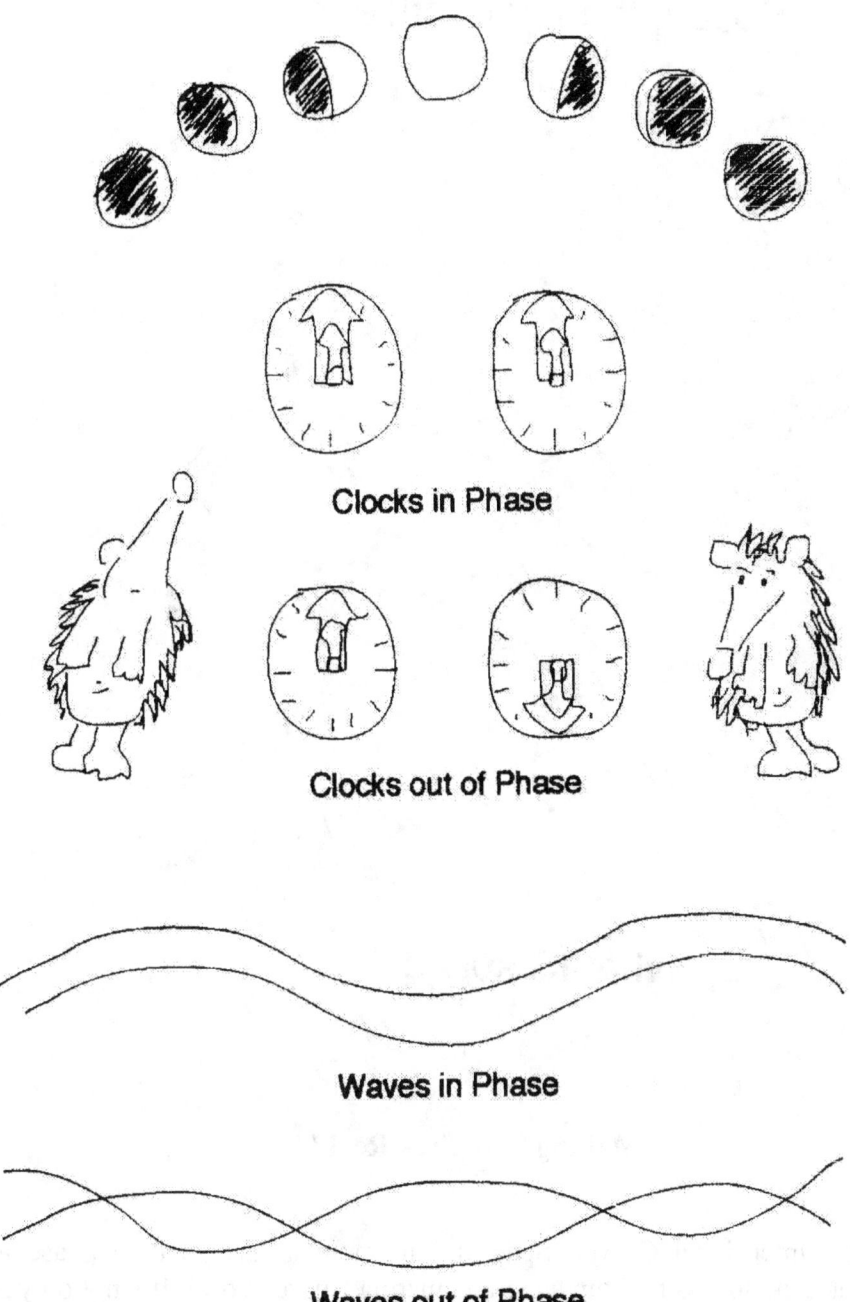

Clocks in Phase

Clocks out of Phase

Waves in Phase

Waves out of Phase

Space, the final frontier?

What IS space?

Making Star Trek Real I

Expanding 3-Dimensional Curved Space is emergent in the vacuum phase transition from random Flat Land to less random superconducting Curve World. It's not only 3D space that stress-energy density curves, but 4D space-time that curves. Think of a four-dimensional "world crystal lattice" that cracks. The fault lines of the cracks form the curvature of space-time according to Hagen Kleinert a physics professor at the Free University of Berlin who worked with Richard Feynman at Cal Tech.

ARTISTS CAN'T DO MATH!

CORRECTIONS!

BY STARMAN JACK

POWERS OF TEN

$$10^0 = 1$$

$$10^1$$

$$10^{-1} = \frac{1}{10^1}$$

$$10^2 = 10 \times 10 = 100$$

$$10^{-2} = \frac{1}{10^2} = \frac{1}{100}$$

$$10^3 = 10 \times 10 \times 10 = 1000$$

$$10^{-3} = \frac{1}{10^3}$$

$$10^4 = 10 \times 10 \times 10 \times 10 = 10,000$$

$$10^{-4} = \frac{1}{10^4}$$

$$10^5 = 10 \times 10 \times 10 \times 10 \times 10 = 100,000$$

$$10^{-33} = \frac{1}{10^{33}}$$

$$10^6 = 10 \times 10 \times 10 \times 10 \times 10 \times 10 = 1,000,000$$

$$10^{-33} cm = L_p$$

$$10^{\log_{10} x} = x$$

$$\log xy = \log x + \log y$$

$$\log \frac{x}{y} = \log x - \log y$$

$$c^2 = a^2 + b^2 \rightarrow ds^2 = g_{\mu\nu} dx^\mu dx^\nu$$

$$\log x^a = a \log x$$

$$e^{\log_e x} = x$$

$$y = e^x \rightarrow x = \log_e y$$

$$\log 0 \rightarrow -\infty$$

$$\log 1 = 0$$

$$y = f(x)$$

$$\frac{dy}{dx} = \underset{\Delta x \to 0}{Lim} \frac{\Delta y}{\Delta x} \equiv \underset{\Delta x \to 0}{Lim} \frac{f(x + \Delta x) - f(x)}{\Delta x}$$

$$\frac{dx^n}{dx} = nx^{n-1}$$

$$\int \frac{dy}{dx} dx \equiv \int \frac{df(x)}{dx} dx = y + C = f(x) + C$$

$$\int_a^b \frac{df(x)}{dx} dx = f(b) - f(a)$$

$$\frac{de^{ax}}{dx} = ae^{ax}$$

$$\frac{d \log x}{dx} = \frac{1}{x}$$

$$\frac{d^n}{dx^n} \equiv \left(\frac{d}{dx} \right)^n$$

$$f(x) \approx f(0) + \frac{df(x)}{dx}\bigg|_{x=0} x + \frac{1}{2} \frac{d^2 f(x)}{dx^2}\bigg|_{x=0} x^2 + \frac{1}{2 \times 3} \frac{d^3 f(x)}{dx^3}\bigg|_{x=0} x^3 + \dots$$

$$= \sum_{n=0}^{\infty} \frac{1}{n!} \left(\frac{d}{dx} \right)^n f(x)_{x=0} x^n$$

$$\sum_{n=0}^{\infty} x^n = \frac{1}{1-x}, 0 < x < 1$$

$$e^{ax} = 1 + ax + \frac{(ax)^2}{2} + \frac{(ax)^3}{3 \times 2} + = \sum_{n=0}^{\infty} \frac{(ax)^n}{n!}$$

$$e^{ix} = \cos x + i \sin x$$

$$i^2 = -1$$

"I've many cheerful facts about the square of the hypotenuse" Major General, "Pirates of Penzance", Gilbert and Sullivan http://www.zeitcom.com/majgen/021msong.html http://people.cornell.edu/pages/wcc3/midi/major-general.mid

Real Dimensions...

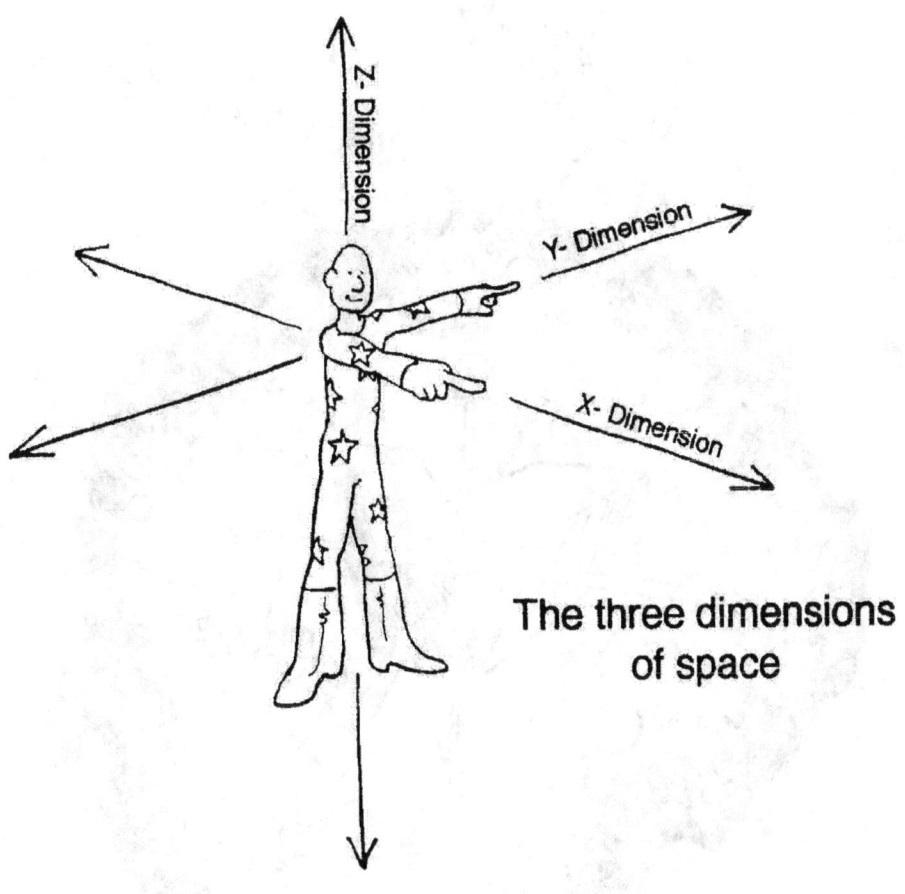

The three dimensions
of space

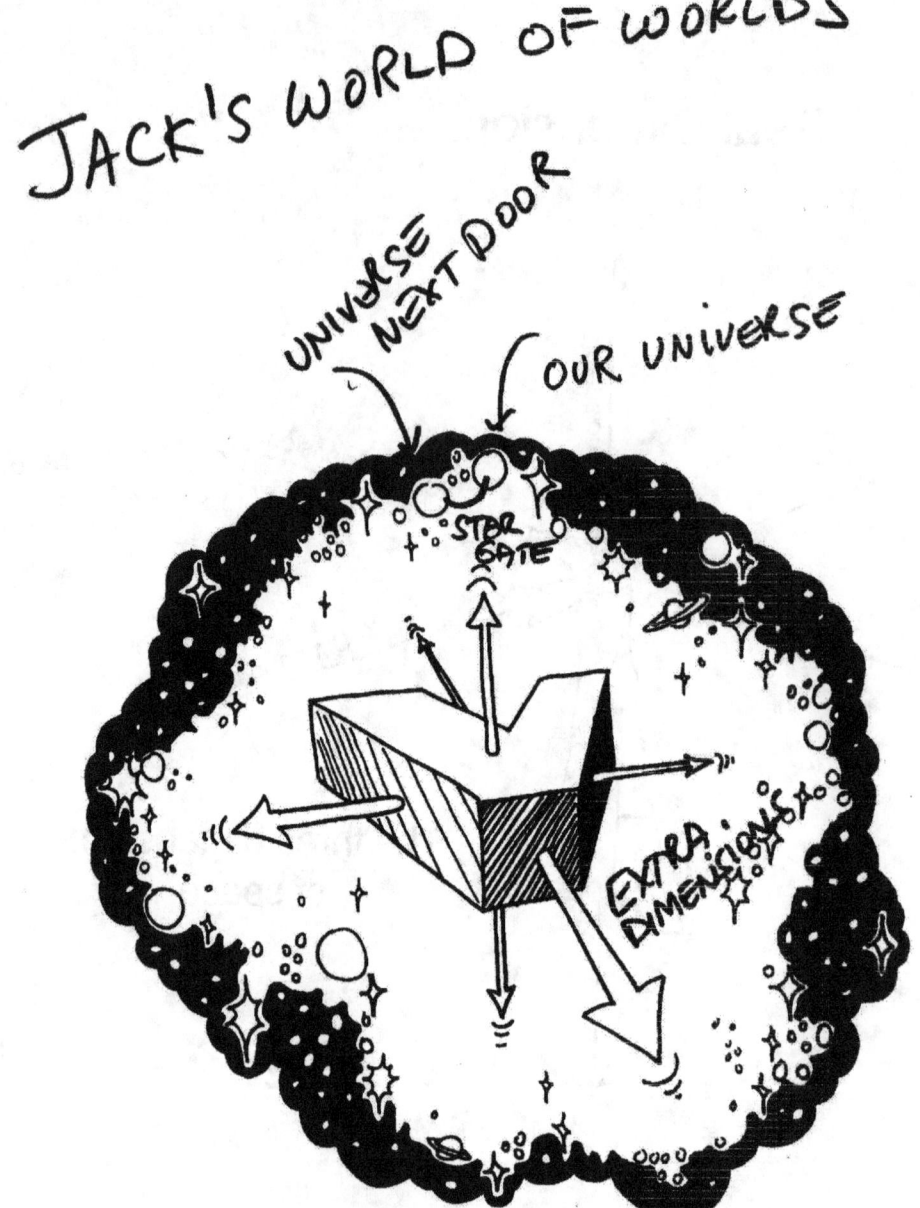

THE BIG BANG

OF OUR UNIVERSE BUBBLE
ONE OF MANY IN SUPERCOSMOS
EVERYTHING OUT OF INITIAL SINGULARITY

OUR UNIVERSE IS NOT IN GLOBAL THERMAL
EQUILIBRIUM BECAUSE OF THE COSMOLOGICAL
EXPANSION OF THREE DIMENSIONAL SPACE

THIS ALLOWS THE CREATION
OF COMPLEX MATTER STRUCTURE
OUT OF INITIAL UNIFORMITY
WITH SMALL DENSITY FLUCTUATIONS
OF $\sim 10^{-5}$, I.e. COBE DATA ON
BLACK BODY TEMPERATURE.

17ᵀᴴ CENTURY
NEWTON'S MECHANICS

$$\vec{F} = 0$$

A BODY IN MOTION TENDS TO STAY IN MOTION

FIRST LAW

A FORCE $\vec{F}$ ACTING ON A BODY WILL CAUSE THAT BODY TO EITHER SPEED UP, SLOW DOWN, OR CHANGE DIRECTION

$$\vec{F} = m\vec{a}$$
SECOND LAW

A FORCE ACTING ON A BODY WILL CAUSE THAT BODY TO RETURN AN EQUAL AND OPPOSITELY DIRECTED FORCE ON THE SOURCE OF THE ORIGINAL FORCE

EARLY ACTION-REACTION
3RD LAW

$$\vec{F}_A = -\vec{F}_B$$

MAXWELL'S ELECTROMAGNETIC FIELD

"THE BOUNDARY OF A BOUNDARY IS ZERO!"

JOHN ARCHIBALD WHEELER

CARTAN FORMS
HOMOLOGY

$$dF = 0$$
$$* d * F = J$$

- CHARGE

WORMHOLE

+ CHARGE

"MASS WITHOUT MASS"

"CHARGE WITHOUT CHARGE"

EINSTEIN'S THEORY OF GRAVITY

SPACE-TIME CURVATURE = SPACE-TIME STIFFNESS
X STRESS-ENERGY
DENSITY

$$G_{\mu\nu} = 8\pi \frac{G}{c^4} T_{\mu\nu}$$

GEOMETRODYNAMICS WITH $\Lambda = 0$

THERMODYNAMICS
OF HEAT AND ENTROPY

FIRST LAW

$$\delta E = T \, \delta S + P \, \delta V$$

ENTROPY CHANGE ↙

VOLUME CHANGE ←

ENERGY CHANGE ↗

ABSOLUTE TEMPERATURE ↑

PRESSURE ↑

SECOND LAW

$$\delta S \geq 0$$

IN CLOSED SYSTEM

ZEROTH LAW

$$T \text{ IS SAME EVERYWHERE}$$

IN THERMAL EQUILIBRIUM

THE BREAKDOWN
OF CLASSICAL PHYSICS

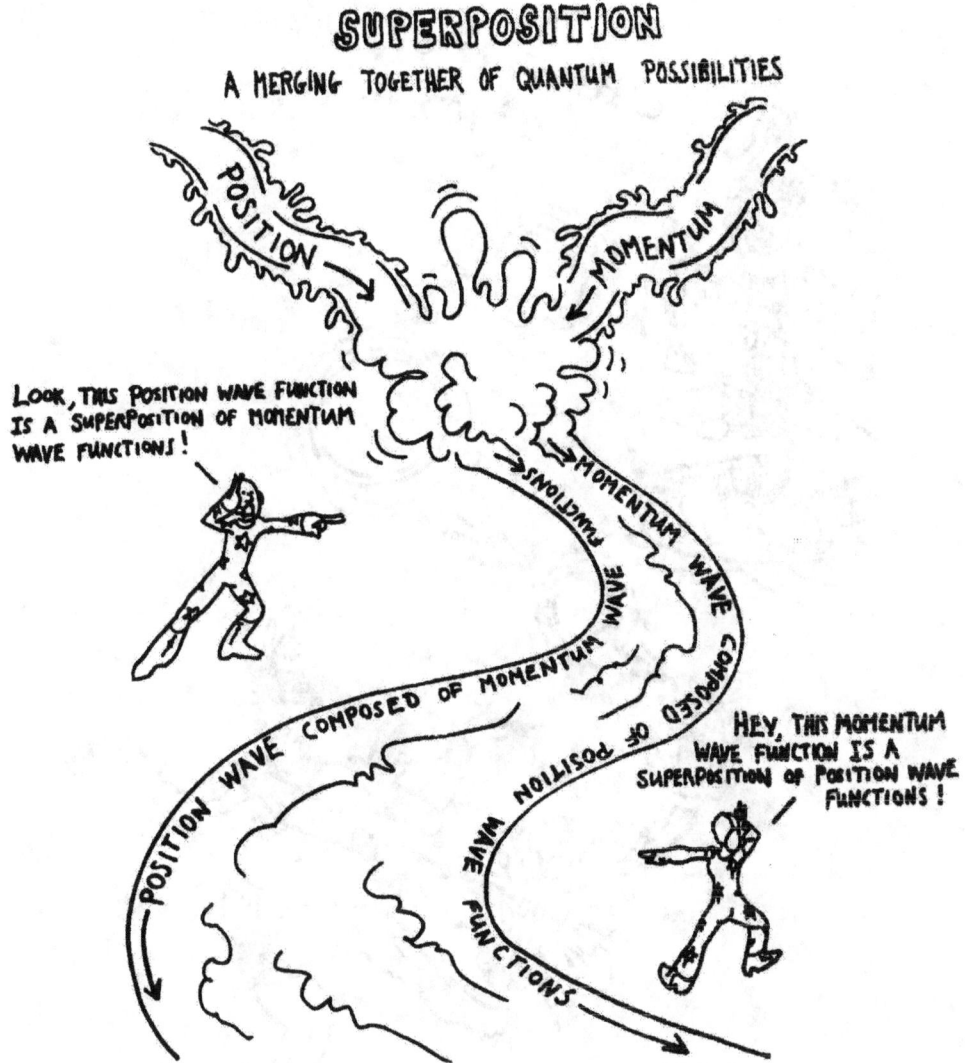

QUANTUM FIELD THEORY

CREATION OF QUANTA
$$a^\dagger |0\rangle = |1\rangle$$

POOF

DESTRUCTION OF QUANTA
$$a|1\rangle = |0\rangle$$

$$|0\rangle = \text{MICRO QUANTUM VACUUM}$$

$$aa^\dagger - a^\dagger a = 1$$

IS BOSON HEISENBERG PRINCIPLE

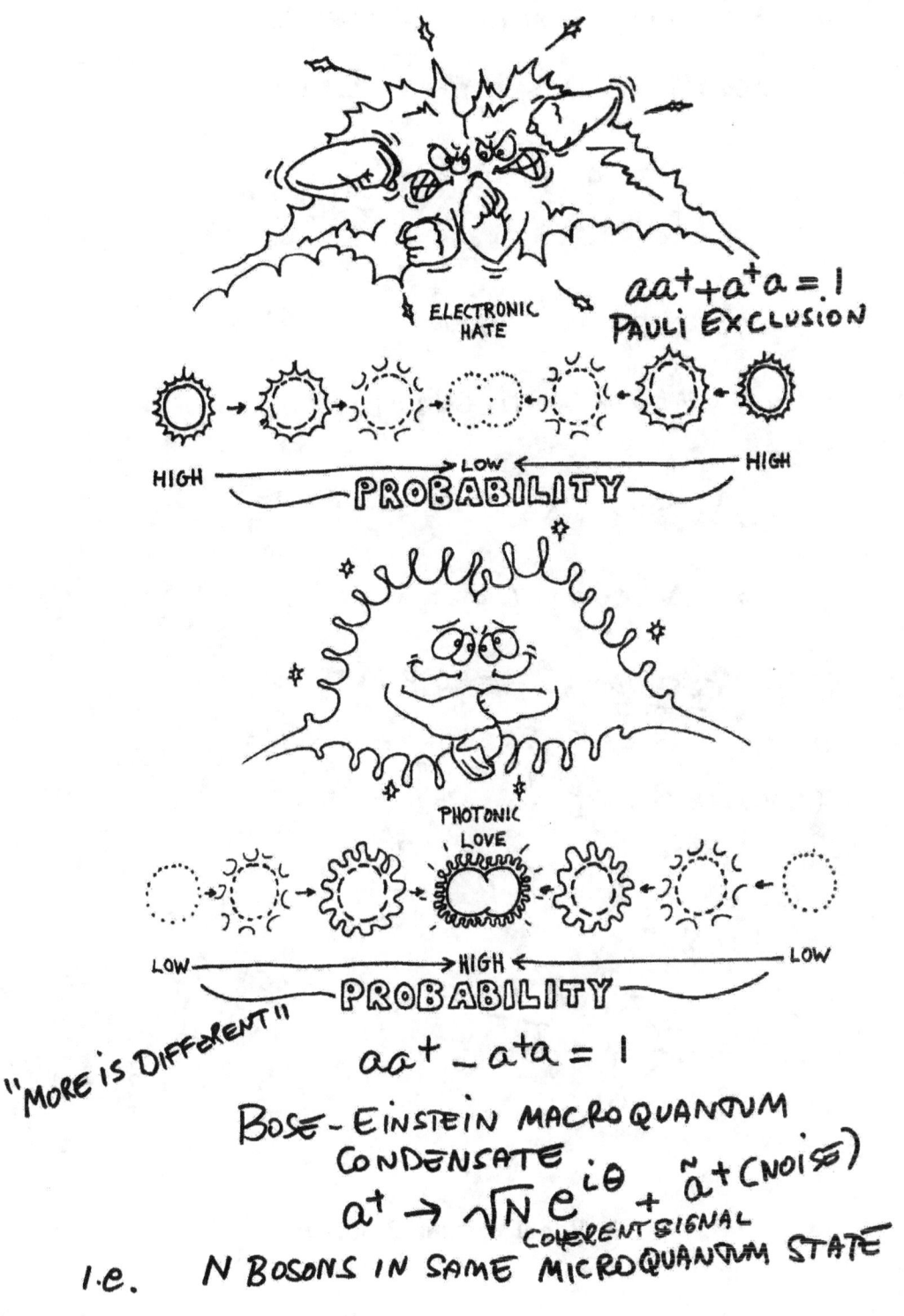

ELECTRONIC
HATE

$$aa^+ + a^+a = 1$$
PAULI EXCLUSION

HIGH ———————→ LOW ←——————— HIGH
PROBABILITY

PHOTONIC
LOVE

LOW ———————→ HIGH ←——————— LOW
PROBABILITY

"MORE IS DIFFERENT"

$$aa^+ - a^+a = 1$$

BOSE - EINSTEIN MACRO QUANTUM
CONDENSATE

$$a^+ \rightarrow \sqrt{N}\, e^{i\theta} + \overset{N}{a}{}^+(\text{NOISE})$$

COHERENT SIGNAL

i.e.   N BOSONS IN SAME MICRO QUANTUM STATE

JACK'S MACRO-QUANTUM GEOMETRODYNAMICS
EINSTEIN'S FIELD EQUATION GETS A NEW TERM

$$G_{\mu\nu} + \Lambda g_{\mu\nu} = 8\pi \left(\frac{G}{c^4}\right) T_{\mu\nu}$$

$\Lambda > 0$ IS "DARK ENERGY"

I.e VACUUM ANTI-GRAVITATES

$\Lambda < 0$ IS "DARK MATTER"

I.e. VACUUM GRAVITATES

Λ IS RESIDUAL ZERO POINT MICRO-QUANTUM HEISENBERG VACUUM NOISE
ALLOWED BY Ψ! $\Psi = |\Psi| e^{i\,PHASE}$

$$\Lambda = \frac{1}{L_p^2} \left(1 - L_p^3 |\Psi|^2\right)$$

$$g_{\mu\nu} = \frac{L_p^2}{2} \left[\frac{\partial^2 PHASE}{\partial x_\mu \partial x_\nu} + \frac{\partial^2 PHASE}{\partial x_\nu \partial x_\mu} \right] + \eta_{\mu\nu}$$

FLAT SPACE-TIME

The Bondi Vacuum Propeller

The flat plate Bondi[57] capacitor vacuum propeller self-propels along a time-like geodesic along the $\Lambda > 0 \rightarrow \Lambda < 0$ direction (to the right in this example). There is no limit to the

---

[57] Herman Bondi, Chief Scientist of The British Ministry of Defense, lectured to us in 1960 at Cornell's Newman Lab for Nuclear Studies on a similar idea using negative mass-energy. This was 42 years before I discovered the unified Λ field, which does what negative mass-energy does plus a lot more!

effective speed reached as seen by outside observers because internally the crew is "not moving at all"! The effective warp speed can be many times c. There are no g-forces and the tidal forces can be kept to a minimum.

Λ-FIELD WARP DRIVE?
SPIN 2 BONDI VACUUM PROPELLER

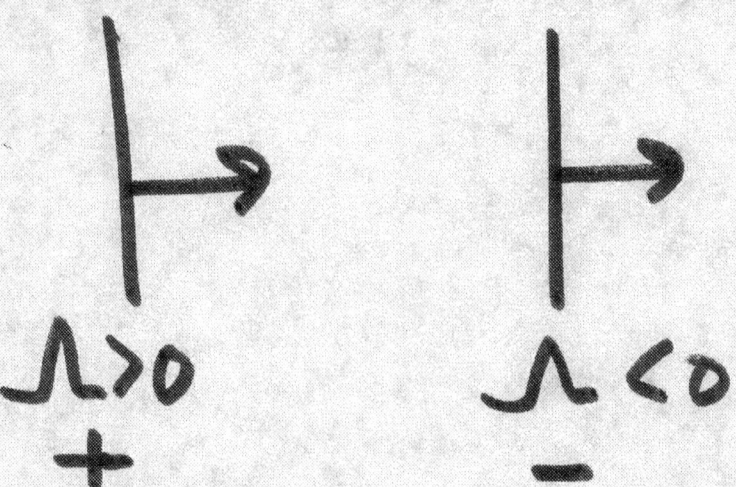

$$\Lambda > 0 \qquad\qquad \Lambda < 0$$
$$+ \qquad\qquad\qquad -$$

CENTER OF MASS ACCELERATES RELATIVE TO OUTSIDE $\Lambda = 0$ WITH NO g-FORCE ON INSIDE $\Lambda = 0$ CREW OF FLYING SAUCER. NO LIMIT ON EFFECTIVE AVERAGE SPEED $> C$ JUST LIKE STAR TREK!

SPIN 1

ELECTRIC CHARGE

$$| \rightarrow \quad \leftarrow |$$

$$+ \qquad -$$

$$\leftarrow | \qquad | \rightarrow$$

$$- \qquad -$$

CENTER OF MASS DOES NOT ACCELERATE

BOHM'S ORTHODOX QUANTUM THEORY

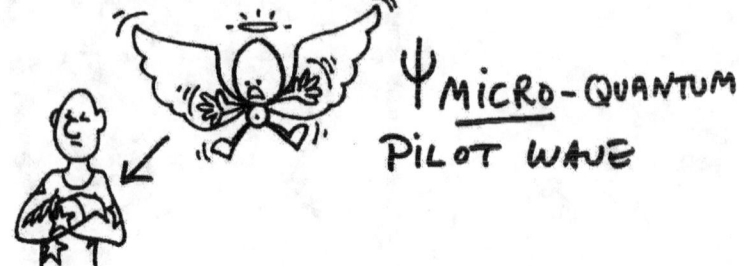

ψ MICRO-QUANTUM
PILOT WAVE

PARTICLE

ψ-ACTION WITHOUT REACTION
SIGNAL-LOCALITY! (NO-CLONES)

JACK'S MACRO POST-QUANTUM THEORY

ψ GIANT LOCAL
MACRO-QUANTUM
PILOT WAVE

"TWO-WAY RELATION" (BOHM-HILEY)

i.e ACTION + REACTION FEEDBACK LOOP
GENERATES CONSCIOUSNESS!
SIGNAL NONLOCALITY!

The quantum field of active information may be compared to
"a ship on automatic pilot being guided by radio waves".

http://stardrive.org/cartoon/bovineshtml

Hidden Variables...

Hidden Variables

We have met the *hidden variables* and they are *our material bodies* piloted by giant, but local, macro-quantum intrinsically mental Ψ_{MACRO} fields of thought. Direct back-action of matter on mind, on its Ψ_{MACRO}-Angel ☺, generates inner feeling as an excited state of Ψ_{MACRO}. In contrast to midget micro-quantum ψ_{micro} fields of photons, nucleons and electrons, giant macro-quantum Ψ_{MACRO} fields do not decohere easily. That is they do not break apart into random unconscious noise. That is what aging and death are – a slow decoherence of our self Ψ_{MACRO} field. We *are* giant Ψ_{MACRO} fields

MICRO-QUANTUM TELEPORTATION? YES!

ALICE — CLASSICAL SIGNAL → BOB ψ OUT CREATE

ψ_{in} DESTROY — EPR PAIR

ψ IS TELEPORTED!

YES, THAT'S RIGHT,
NONLOCAL CONNECTIONS ARE
INSTANTANEOUS!

UNTAPPABLE BY EVE? YES!
CLONE A QUANTUM? NO!
i.e MICRO QUANTUM NONLOCALITY YET SIGNAL LOCALITY!
BORN PROBABILITY ~ $|\psi|^2$
PROBABILITY IDEA BREAKS DOWN IN $\psi_{micro.} \Rightarrow \psi_{macro}$
HENCE MACRO QUANTUM SIGNAL NONLOCALITY
ie LIBET, RADIN, BIERMAN "PRESPONSE".

It's a good thing we screened out Decoherence,
or we'd never get any backactivity done!

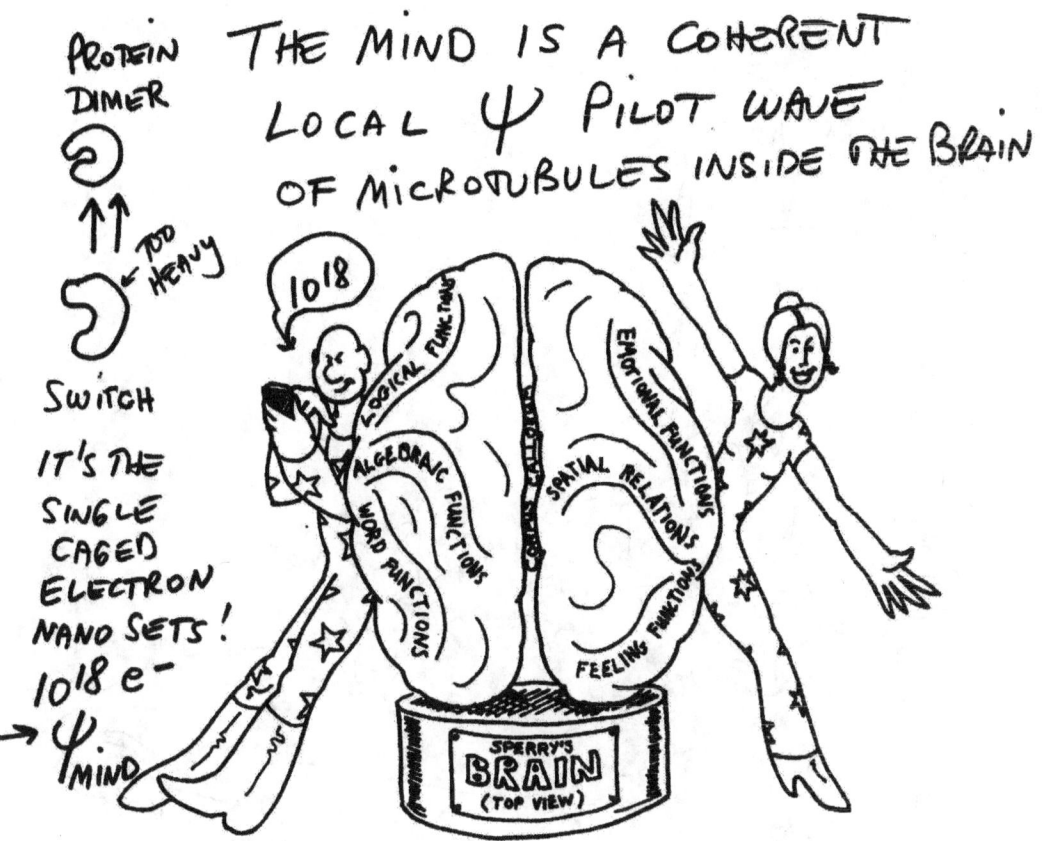

THE MIND IS A COHERENT LOCAL ψ PILOT WAVE OF MICROTUBULES INSIDE THE BRAIN

THE ψ MIND FIELD IS
MACRO QUANTUM. THEREFORE
ITS "PHASE RIGIDITY"
RESISTS DECOHERENCE.
e.g. PW ANDERSON'S "MORE IS DIFFERENT"

85

Run, they've achieved catalytic closure!

Feel like a Neanderthal around your peers?

Try a Bio-RAM nano-quantum chip intellect upgrade-- after all,
staying dumb is just plain stupid.

BLACK HOLE THERMO

SECOND LAW

$$\delta A \geq 0 \rightarrow \delta(S + cA) \geq 0$$

$A =$ SURFACE AREA OF BLACK HOLE
WHERE TIME STOPS!

HAS $\dfrac{A}{L_p^2}$ BITS OF INFORMATION

$L_p^2 =$ PLANCK AREA $= \dfrac{hG}{c^3} = 10^{-66} \, cm^2$

$h =$ PLANCK'S QUANTUM
OF ψ - ACTION ON MATTER

$G =$ NEWTON'S GRAVITY

$c =$ SPEED OF LIGHT IN VACUUM

FIRST LAW

↓ ROTATION

$$\delta E = \frac{k}{8\pi} \, \delta A + \Omega \, \delta J + \Phi \, \delta Q$$

↗ ELECTRIC CHARGE

ZEROTH LAW
SURFACE GRAVITY $k =$ CONSTANT ON BLACK HOLE
SURFACE A.K.A HORIZON

DESTINY MATRIX

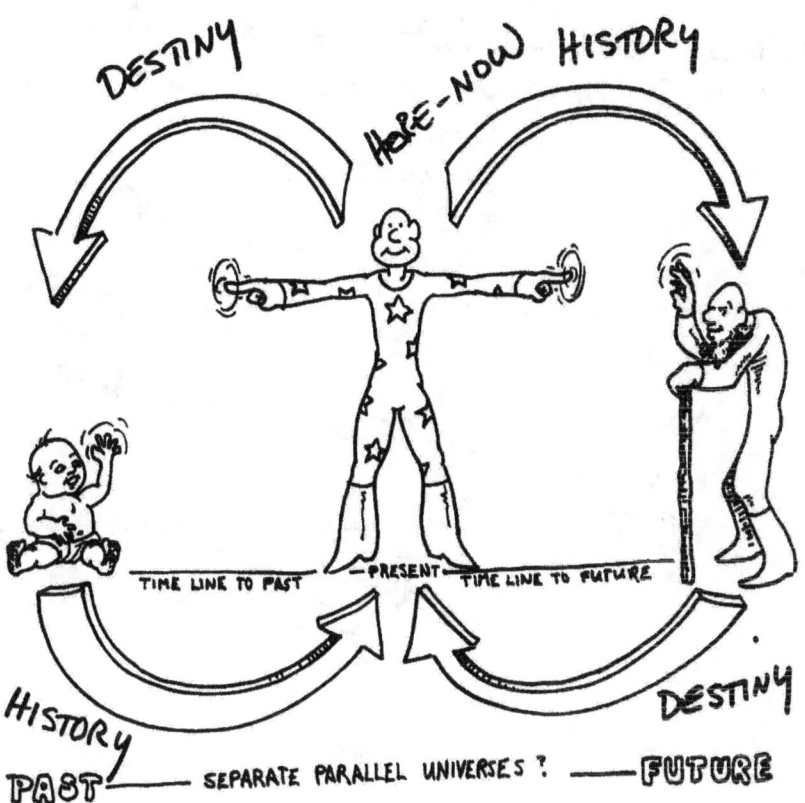

THE DOUBLE TIME LOOP!

SEE ALSO FREDDY'S PAPER ON LIBET'S
BRAIN-MIND PRESPONSE
AND AHARONOV'S TWO-VECTOR QUANTUM THEORY.

METRIC ENGINEERING 101

PUTHOFF'S PV? NO!

HAISCH'S ZERO-POINT INERTIA? NO!

NICK COOK'S "HUNTFOR THE ZERO-POINT"? NO!

U.F.O.s:

Do they exist?

$$\mathbf{G}_{\mu\nu}{}^{;\mu} + \Lambda^{,\mu} g_{\mu\nu} = 8\pi \frac{G}{c^4} T_{\mu\nu}{}^{;\mu}$$

IS EQUATION OF WEIGHTLESS <u>WARP DRIVE</u>

i.e. "VACUUM PROPELLER"

; = COVARIANT DERIVATIVE , , ORDINARY PARTIAL DERIVATIVE

CABALA! "TRADITION!" ZERO MOSTEL
"FIDDLER ON THE ROOF"!

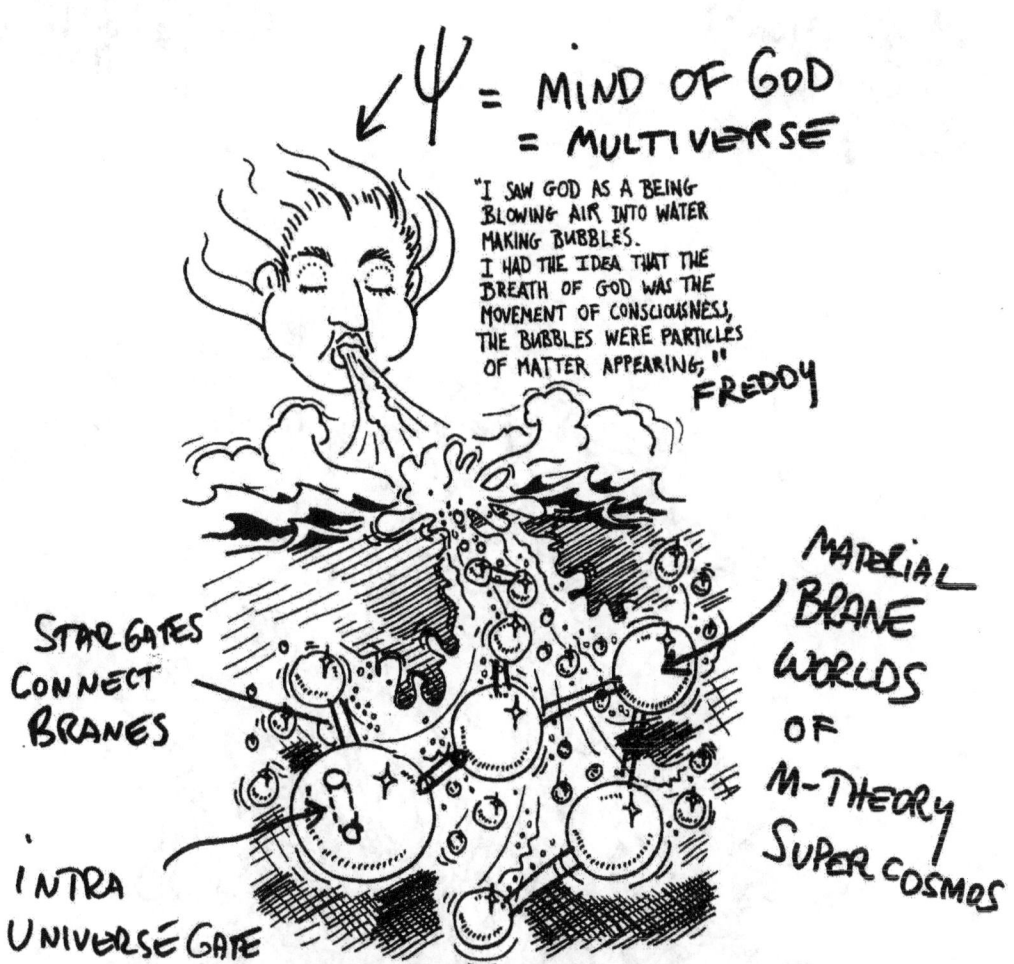

Ψ = MIND OF GOD
= MULTIVERSE

"I SAW GOD AS A BEING
BLOWING AIR INTO WATER
MAKING BUBBLES.
I HAD THE IDEA THAT THE
BREATH OF GOD WAS THE
MOVEMENT OF CONSCIOUSNESS,
THE BUBBLES WERE PARTICLES
OF MATTER APPEARING,"
FREDDY

MATERIAL
BRANE
WORLDS
OF
M-THEORY
SUPER COSMOS

STARGATES
CONNECT
BRANES

INTRA
UNIVERSE GATE

AND GOD WAS THE SPIRIT. THUS NOT ONLY DID GOD
BLOW CONSCIOUSNESS INTO THE FORM OF MATTER, BUT
SIMULTANEOUSLY THIS ACTION WAS THE AWARENESS OF THOSE
BUBBLES AS MATTER." THE EAGLE'S QUEST p. 43

" CIPHER OF GENESIS "
- CARLO SUARES
"MY FATHER'S HOUSE HAS MANY MANSIONS."
RABBI YESHUA

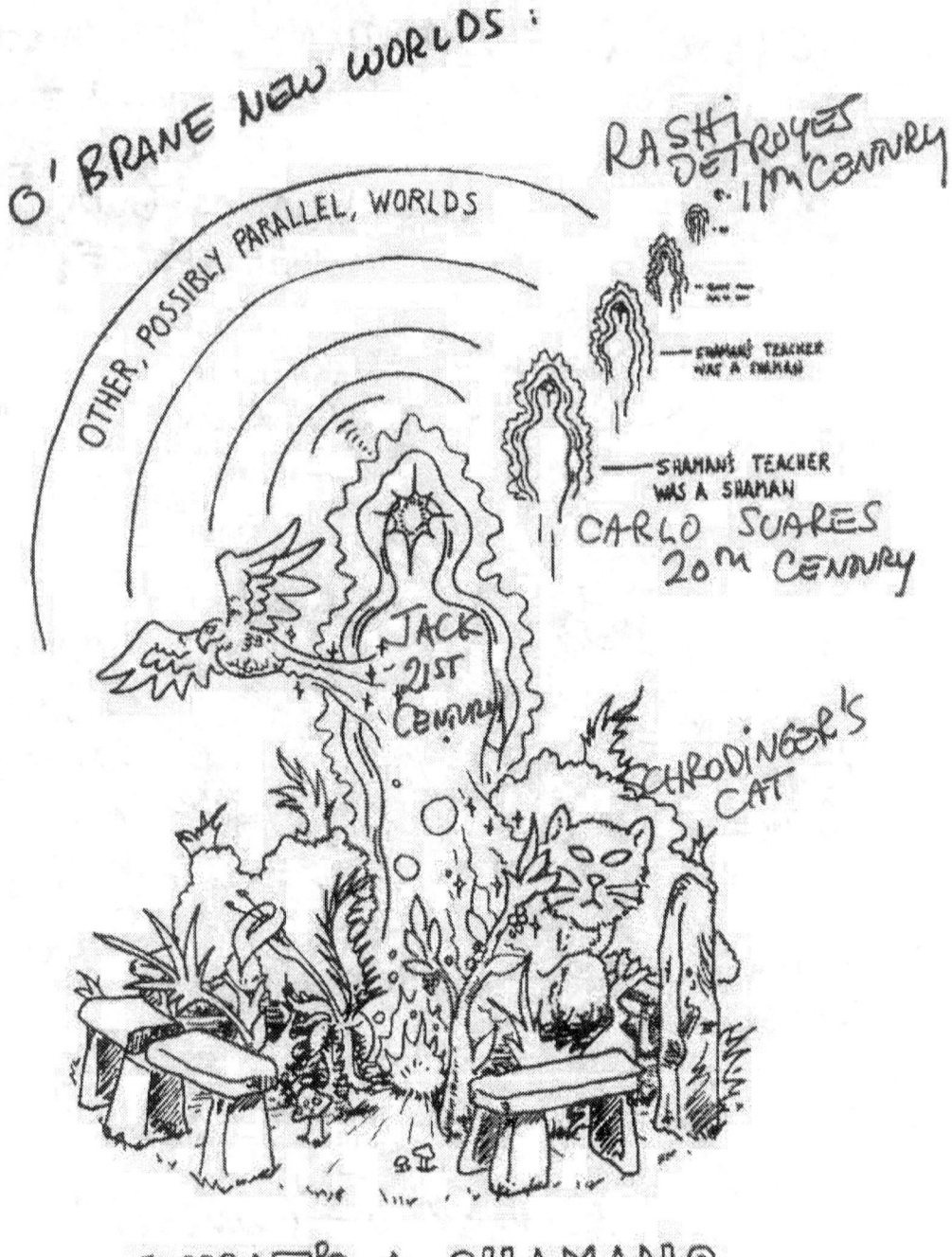

Conversations between Marshall Naify and Jack Sarfatti in the late 1990s

Marshall unfortunately passed away before our project could be completed. I am fulfilling his request to me.

At first Marshall Naify did not want me to directly mention and involve him in the daily exchanges on the Internet. I have since clarified this with him. He no longer has the same objections to that because he understands now that one can't really divorce oneself from this revolutionary process in global communication. We have been quite involved meeting frequently since the first of the year. In the earlier part it was my explaining to him and suggesting that he read certain physics books. I was surprised that Marshall could pick up the essential ideas of the physics theories of the past and present so quickly. Marshall quickly evolved from simply searching for information from Stapp, Bohr, Heisenberg, Einstein and Bohm to reformulating them in interesting ways. Marshall immediately focused on the apparent conflict between relativity and quantum theory. He did not see how the two were incompatible in the overall concept of science. We talked about how to use mathematics to substantiate this position. This apparent conflict of mutually exclusive positions and how to reconcile them will be a major topic. We were and are still involved in developing new insights into the deep issues that interest every thoughtful person in today's fast-changing world. Now that he is more familiar with the instant exchange of ideas on the Internet and has read some of the posts he wants to be in the loop.

A note on Marshall Naify: Marshall Naify is the retired chairman of United Artists Communications and is presently co-chairman of the Todd AO Corporation. His background covers a rather thorough study of history, philosophy, metaphysics, theology emphasizing the Eastern teachings of alchemy, astrology and other esoteric studies. He has recently delved into the present state of theoretical physics and the mind-matter problem.

Marshall: "All things consist of body, mind, and spirit. The brain is the instrument of mind. Mind divides into objective mind and subjective mind, although they both share the same brain function and mechanism. They are complementary. The tempo of your body quiets down to a state where the subjective mind comes into play. The laws of the subjective mind are different from the laws of the objective mind. Time is different. Objective mind encompasses time's flow as is commonly understood in terms of past, present and future. When one meditates you are in the subjective mind, which you are not in when you are awake. You escape the space and time constraints of objective mind. You go immediately to the past, present or to the future or to any time and to any place (location). Material walls and other obstacles are not barriers in the workings of subjective mind. Masters of the Eastern Mystery Schools know these ideas. My understanding of this phenomenon of subjective mind is not only a matter of knowing or learning, but is also from the practical experience of doing — and it's called the art of meditation. That is why people spend a great deal of effort to learn how to meditate. It's not easy to turn off the flow of sense perceptions that feeds the objective mind. We will use the Internet to communicate this understanding to those who are ready to receive it, to enable them to see in a way that will add to objective knowledge. I am not trying to change the present understanding of objective knowledge, but to add to it. Subjective mind is complementary to objective mind and will add fullness to it. We will never know all there is to know. There is always more. It is a bottomless cup. This will appeal to only a small number of people out of the total number that we reach. It was and is the select few, with a strong enough desire to know and the will to pursue an elusive goal. Some people will intellectually understand it, but not experience it. (meditation) because it must be a persistent discipline that is constantly practiced."

Jack: This is like Bohr and Wheeler's characterization of the difference between the classical observer standing outside of the phenomenon, and the quantum observer-participator that becomes an inseparable part of the phenomenon actively shaping it. Also your idea of the fullness of subjective mind as a bottomless cup reminds me of Godel's incompleteness theorem which says that all great mathematical theories have important true equations that cannot be solved within the theory, but require a still greater theory. For example, algebraic equations using only real numbers can have solutions or "roots" that are not real numbers. One must add new imaginary numbers to the theory of real numbers. As you often say, the idea of "fullness" points to the inexhaustibility of the "Divine Mind" or what some physicists call "The Mind of God".

Marshall: "Yes, so far, so good, it is definitive only until there is further delineation and tomorrow will bring new insights. I have begun to reflect on Stapp's writings from my own point of view. Now tell me more about the present state of physics and the role of mind in it. I want to focus on these problems. Of course, Stapp's *Matter, Mind and Quantum Mechanics* outlined for me many of the leading physicists and their theories. I certainly don't agree with his conclusions. We started this conversation twenty years ago Jack, in North Beach in San Francisco in the 1970's. You were different, I was different, the times were different. What I am trying to say is that my own ideas have evolved since those early days and they incorporate advances by many outstanding people that has led me to this present thinking. Our schools didn't teach us about the subjective mind, and still don't. It's time that some attention is paid in this direction, which is so vital to the understanding of brain-mind. It's devastating to think that half of the great faculty of consciousness is neglected. We will never fully understand a thing if we only know it objectively. Knowing something objectively is only knowing a part of it. Knowing about it subjectively complements the objective and gives one a fullness of knowing that may be had in no other way."

C wrote: Yes, but quantum reality is an aspect of physical reality. Indeed it is the basis for physical reality. The pilot wave, from the Bohm picture, is a physical object.

Marshall Naify: "What does C mean by "quantum reality is an aspect of physical reality?" Let me comment on the macrocosm and the microcosm in my realm of how I think of mind. The macrocosm would be "the great Above" or "the great mind of the creator" everything we have here from the beginning of time to the present is taken from "the Mind of God". In other words, it's a mental process. It is not only "As above, so below", but also "As below, so above". Bohm used "As above, so below", but left out "As below, so above". Let me highlight and explain that. How it works is this: take any subject like the theory of learning or education. It's all up there. Everything that is down here has its counterpart above. That means that everything that happens down here has a function up there, but with a difference. With God wanting something becomes synonymous with having it. The way it is, all the subjects here are covered there. But we are not advanced enough to see them in their fullness. One can travel instantly up there while it takes us time to travel down here. Exactly how the subjects in the Above make the shadows down here is hidden from us. Saint Paul said "we see through the glass darkly". What we have down here is an aspect of that. Physical reality down here is an aspect of the "Above". The Eastern Masters said "As below, so above". I am only relating the Great Teaching. The world is a physical reality and an aspect of the spiritual conception, just as in Plato's Cave where all they could see was the shadows moving. That which makes the shadows is the spiritual whole which has mentalized it.

The mind of the creator creates the thought, which becomes the physical reality.

The important numbers are zero, one, two and three. Three is the marriage of one and two. One and two are different. One is masculine. Two is feminine. Three is the Son of one and two. Zero, on the other hand ... I see a missing component of most dual equations as having a third part that is left out. It is a linear extension. It has to do with the circle. Any line that is the least bit curved turns into the circle. It's then a question of how big a circle it is. Confine our discussion to the linear. When I look at the Sun I see straight lines as well as the particle and the wave. Here's what is interesting about zero. Theories are not complete in a dualistic state. The third component is linear. Do analogy with coin. Edge on it's a line. Your subjective mind can see the other side, which the objective mind cannot see directly. The two faces of the coin symbolize the duality. The edge on view is the third part. So duality is as aspect of a triune principle. Water is the medium of mind. The moonlight is a reflection of sunlight. In the ancient Tarot, the moon is a symbol of wisdom and mind. It's like getting the light refracted. It's attuned to the subjective mind."

SCIENTIFIC COMMENTARY

Dr. JACK SARFATTI

Advanced Physics Technical Appendices

Dark Energy

Frank Wilczek wrote[58]:

"Any theory of gravity that fails to explain why our richly structured vacuum, full of symmetry breaking condensates and virtual particles, does not weigh much more than it does is a profoundly incomplete theory."

I think I understand this problem in a unified conservative parsimonious way that is in accord with the experimental data presented in your "Search And Discovery"[59]. According to Bertram Schwarzschild:

"The Boomerang, Maxima, and DASI fits to the CMB power spectrum appeared to confirm an astonishing finding of earlier supernova surveys ... namely, that Ω_Λ is about twice as big as Ω_m . In other words, mass plays second fiddle to a "dark" vacuum energy, with the result that the Hubble expansion is actually speeding up in the present epoch. The familiar world is further demeaned by the realization that Ω_b, the cosmic density of ordinary baryonic matter, constitutes only about 15% of Ω_m, the rest being some sort of exotic matter only slightly less mystifying than the dark energy."

My basic idea here is that the smooth nonrandom coherent symmetry breaking virtual *macro-quantum* vacuum Bose-Einstein condensates *compensate* the *micro-quantum* random incoherent residual virtual particle zero-point vacuum fluctuations of Heisenberg uncertainty noise. The virtual condensates can be pictured as complex numbered coherent signals $\Psi \equiv |\Psi| e^{i \arg \Psi}$ whose phase variations yield Einstein's classical geometrodynamic field $g_{\mu\nu}$ of curved space-time, and whose amplitude variations damp down the "quintessent" Λ field from the virtual particles forming the Heisenberg uncertainty zero-point random noise.

---

[58] "Scaling Mount Planck III: Is That All There Is?", p. 11, Physics Today, August, 2002
[59] "Observing the Cosmic Microwave Background at High Resolution Bolsters the Inflationary Big Bang Scenario", p. 20, Physics Today, August, 2002

Virtual bosons have positive random incoherent zero-point energy density with equal negative zero-point pressure.[60] The contribution of the zero-point pressure is three times stronger than the zero-point energy density. Therefore, the net result of virtual off-mass shell zero-point quanta is gravitationally repulsive making a huge unobserved positive contribution to the Λ field of order $L_p^{-2} \sim c^3/hG \sim 10^{66}$ cm^{-2}, the reciprocal Planck area.[61] Therefore a dominance of virtual bosons on the cosmological scale makes the universe accelerate because the "dark energy" vacuum at that scale literally would anti-gravitate. Note my old Cornell Chum Lenny Susskind's conjecture of "UV/IR duality"[62] in which the smallest of scales meets the largest of scales. The problem, as Wilczek alludes to, is that this naïve estimate is about 122 powers of ten too big! This is because the actual net Λ is very close to zero not much more than the reciprocal square of the Hubble radius of 10^{28} cm.

Virtual *unbound* [63] fermion-antifermion pairs[64] of the micro-quantum polarized vacuum have the opposite zero-point effect.[65] Therefore, purely random virtual fermion-antifermion pairs with positive dominating zero-point pressure make a negative contribution to Λ on somewhat smaller scales than cosmological corresponding to a gravitating phase of vacuum misnamed the "dark matter"[66] that is $\sim 85\%$ of Ω_m according to CBI et-al.

Einstein's gravity with quintessence as emergent macro-quantum vacuum order

The role of the virtual condensates is to modulate the positive anti-gravitating contributions to the local variable vacuum Λ-field from all the virtual zero-point gauge force bosons and the negative gravitating contributions from all the virtual zero-point gauge source unbound fermion-antifermion pairs. Consider only quantum electrodynamics for simplicity. The fuzzy edge of the Dirac-Fermi surface vacuum of filled negative energy electron states is unstable to the BCS type pair formation[67] from the intrinsic Coulomb attraction between virtual electrons and virtual positrons. Therefore, we expect a Bose-Einstein condensation into a macroscopically occupied virtual bound state with local *scale-*

---

[60] "Cosmological Physics", eqs. (1.85), (1.86) p. 25 & (1.87), (1.88) p. 26, John Peacock, Cambridge, 1999

[61] Each Planck area is worth 1 c-bit of Bekenstein-Hawking gravitational entropy per Boltzmann's constant not from the usual coarse-graining of statistical mechanics. Lenny Susskind conjectures that the number of c-bits in any volume V of 3D space is $\sim V^{2/3}$/Planck Area with The World as a Hologram.

[62] "Twenty Years of Debate With Stephen", hep-th/0204027

[63] Virtual ionized neutral plasma.

[64] Zero-point vacuum polarization

[65] "The Quantum Vacuum", 10.6 eqs. (10.105) (10.106) pp 353-354, Peter W. Milonni, Academic Press, 1994

[66] Schwarzschild calls this $\Lambda < 0$ vacuum phase "exotic matter", but that should not be confused with the $\Lambda > 0$ "exotic matter" (AKA "dark energy") Kip Thorne needs to make a traversable wormhole.

[67] We do not need the electron-phonon interaction here as in a real superconductor with real electron-electron pairs. The Fermi momentum is $\sim h/L_p$. The binding energy of the virtual pair is $\sim \alpha m_p c^2 \sim -10^{17}$Gev $\sim$ critical temperature to destroy vacuum superconductivity. The condensation energy density is $\sim -L_p^{-3}(m_e/m_p)\alpha\, m_p c^2 \sim$ $\sim -L_p^{-3}\alpha\, m_e c^2 \sim 10^{99}\, 10^{-2}$ Mev/cc. The photon rest mass is $\sim 10^{-65}$ gm with a Meissner penetration depth $\sim 10^{28}$ cm, so that the ratio of penetration depth to coherence length of the macro-quantum vacuum $\gg 1$, i.e. a hard superconductors with magnetic vortex string topological defects.

dependent complex macro-coherent order parameter $\Psi(x,\sigma)$.[68] Einstein's local geometrodynamic field is[69]

$$g_{\mu\nu}(x,\sigma) = \eta_{\mu\nu} + \frac{L_p^2}{2}\left[\partial_\mu\partial_\nu + \partial_\nu\partial_\mu\right]\arg\Psi(x,\sigma) \tag{1.1}$$

Where $\eta_{\mu\nu}$ is the flat space-time Minkowski metric tensor. The second term on the RHS of (1.1) need not be small. This is not a linear perturbation theory. I extend Einstein's local field equation to

$$G_{\mu\nu}(x,\sigma) + \Lambda(x,\sigma)g_{\mu\nu}(x,\sigma) = -8\pi\frac{G}{c^4}T_{\mu\nu}(x,\sigma) \tag{1.2}$$

With the covariant Landau-Ginzburg equation[70]

$$D^\mu D_\mu \Psi(x,\sigma) + \alpha\Psi(x,\sigma) + \beta\left|\Psi(x,\sigma)\right|^2 \Psi(x,\sigma) = 0 \tag{1.3}$$

"More is Different" broken "Goldstone" symmetry needs $\alpha < 0$, $\beta > 0$. I introduce a "wavelet transform" scale parameter σ[71]. Furthermore, this virtual "two fluid" model of the vacuum has

$$\Lambda(x,\sigma) = \frac{1}{L_p^2}\left[1 - L_p^3\left|\Psi(x,\sigma)\right|^2\right] \tag{1.4}$$

Conservation of "momenergy"[72] generalizes to

$$D^\nu G_{\mu\nu} + g^{\nu\lambda}\partial_\lambda\Lambda g_{\mu\nu} = -8\pi\frac{G}{c^4}D^\nu T_{\mu\nu}$$

My new virtual superfluid vacuum theory of the variable quintessent $\Lambda(x,\sigma)$ field explaining both "dark energy" and "dark matter" as positive and negative values, respectively, of the same field at different scales σ may be equivalent to a torsion field theory.[73] I *conjecture* that

---

[68] "A Career in Theoretical Physics", P.W. Anderson, 1994 "More is Different", "Coherent Matter Field Phenomena in Superfluids", "Macroscopic Coherence and Superfluidity", "Hard Superconductor" et-al.

[69] This is Wheeler's "IT FROM BIT", i.e. GEOMETRY from INFORMATION.

[70] "Hard Superconductors" P.W. Anderson. This is "BIT FROM IT", i.e. INFORMATION from GEOMETRY illustrating no action without reaction like in Escher's "Drawing Hands". This extra detail is not in Wheeler's writings as far as I know.

[71] A Friendly Guide to Wavelets", Gerald Kaiser, Birkhauser, 1994. The wavelet transform with adaptive window scale σ is more suited to variably curved space-time where the rigid Fourier transform definition of the Wigner phase space density breaks down. The macro-quantum order parameter $\Psi(x,\sigma)$ is related to a Wigner scale space density generalization of the Wigner phase space density. "Phase Space Picture of Quantum Mechanics", Y.S. Kim, M.E. Noz, World Scientific, 1991

[72] "A Journey into Gravity and Space-time", Ch 6, John Archibald Wheeler, Scientific American, 1990

[73] "A Theory of Physical Vacuum", Gennady Shipov, Moscow, 1998 (ISBN 5-7273-0011-8)

$$g^{v\lambda}\partial_{\lambda}\Lambda \overset{?}{=} T_{\lambda}^{v\lambda} \qquad (1.5)$$

Where $T_{\mu}^{v\lambda} = -T_{\mu}^{\lambda v}$ is the 3rd rank Shipov "contortion tensor".

The usual Bianchi identity result

$$D^{v}G_{\mu v} = 0 \qquad (1.6)$$

only works when the manifold has zero torsion.[74]

The Macro-Quantum Einstein Gravity Action
The general relativity Einstein-Hilbert field Lagrangian for curvature scalar R coupled to a spin zero complex scalar field is

$$L = \frac{c^4}{G}\int R\sqrt{g}\,d^3x + \int\sqrt{g}\left[g^{\mu v}\left(\hbar c\right)^2\partial_{\mu}\varphi^*\partial_{v}\varphi - m^2c^4\varphi^*\varphi + \beta\left(\varphi^*\varphi\right)^2\right]d^3x \qquad (1.7)$$

Where the physical dimensions of the spin 0 field are:

$$[\varphi] = \frac{1}{\sqrt{EL^3}} \qquad (1.8)$$

Note that

$$g \equiv \det g_{\mu v} \qquad (1.9)$$

Since the macro-quantum vacuum is superfluid, as in the two fluid model with ODLRO

$$\varphi = \langle\varphi\rangle + \hat{\varphi}$$
$$\Psi = \sqrt{E}\langle\varphi\rangle \qquad (1.10)$$

The second quantized "normal fluid" operator $\hat{\varphi}$ destroys an unbound virtual electron-positron pair in the random plasma near the fuzzy Heisenberg edge of width $\sim 2m_ec$ of the Dirac-Fermi sphere of radius h/L_p of the micro-quantum electron vacuum and adds a bound virtual positronium to the Bose-Einstein condensate. The Hermitian conjugate $\hat{\varphi}$ ionizes a virtual pair out of the coherent condensate into the random noisy plasma but still inside the

---

[74] Wheeler and Ciufolini, "Gravitation and Inertia". See also Hagen Kleinert "Gauge Fields in Condensed Matter, Vol II Stresses and Defects" in which curvature and torsion are infra-red disclination and dislocation topological defects in the Planck-scale 4-D "world crystal lattice". I relate this to phase singularities in eq. 1.1, which generalizes the deBroglie-Bohm pilot wave guidance constraint from quantum fluid flow to quantum relativistic crystal elasticity. Eq 1.1 can be made gauge invariant in the usual minimal coupling way as in the micro-quantum Bohm-Aharonov effect and the macro-quantum Josephson effect.

two-fluid vacuum of course. The effective boson is the center of mass motion of the virtual electron-hole (positron) pairs. The relative motion is integrated out.

The effective Landau-Ginzburg Bose-Einstein condensate potential energy *density* with spontaneous broken vacuum symmetry is

$$V_{eff} = -m^2 c^4 \varphi^* \varphi + \beta (\varphi^* \varphi)^2 \qquad (1.11)$$

$$\frac{\delta V_{eff}}{\delta \varphi^*} = -m^2 c^4 \langle \varphi \rangle + 2\beta \langle \varphi \rangle^* \langle \varphi \rangle^2 = 0$$

$$\langle \varphi \rangle \neq 0 \qquad (1.12)$$

$$|\langle \varphi \rangle|^2 = \frac{m^2 c^4}{2\beta}$$

$$V_{eff\,min} = -\frac{(m^2 c^4)^2}{2\beta} + \frac{(m^2 c^4)^2}{4\beta} = -\frac{(m^2 c^4)^2}{4\beta} \qquad (1.13)$$

This condensate term adds an attractive "dark matter" contribution to R in the Hilbert gravity Lagrangian in the form of a variable cosmological field, i.e. $R - 2\Lambda$ where

$$\Lambda_{bec}(x) = \frac{G}{c^4} V_{eff\,min}(x) = -\frac{G}{c^4} \frac{(m^2 c^4)^2}{4\beta(x)} = -\frac{G}{c^4}(mc^2)^2 |\langle \varphi(x) \rangle|^2 = -Gm_p^2 |\langle \varphi(x) \rangle|^2 \quad (1.14)$$

That is, the field Lagrangian is now

$$L = \frac{c^4}{G} \int (R - 2\Lambda_{bec}) \sqrt{g}\, d^3 x + \int \sqrt{g} \left[\begin{array}{c} g^{\mu\nu}(\hbar c)^2 \partial_\mu (\langle \varphi \rangle + \hat{\varphi})^* \partial_\nu (\langle \varphi \rangle + \hat{\varphi}) \\ -m^2 c^4 [\langle \varphi^* \rangle \hat{\varphi} + \langle \varphi \rangle \hat{\varphi}^* + \hat{\varphi}\hat{\varphi}^*] + \beta(....)^2 \end{array} \right] d^3 x \quad (1.15)$$

The second part of the RHS of (1.15) has the graviton-condensate-virtual electron-positron plasma interactions.

The random zero point noise of the effective virtual bosons adds a positive antigravitating "dark energy" contribution of order L_p^{-2} to the cosmological field as do the virtual zero point spin 1 photons. The net result corresponding to (1.4) is

$$\Lambda_{eff} = \Lambda_{zpf} - \Lambda_{bec} \approx \frac{\varsigma}{L_p^2} - -Gm_p^2 |\langle \varphi(x) \rangle|^2 \qquad (1.16)$$

$$Gm_p^2 = \hbar c$$

The dark energy-matter crisis in physics today

"dark matter alone is not abundant enough to halt the observed universal expansion. More surprising still is the suggestion that the dominant energy in the universe may be associated with empty space, and moreover, that this energy is causing the expansion of the universe to accelerate with time. Exotic dark matter dominating the mass of the universe is a concept that may be hard for some to accept. But at least dark matter is 'stuff', however exotic. How much harder is it to get used to the idea that most of the energy density of the universe may literally be associated with nothing at all." Preface to "Quintessence" by Lawrence Krauss (Basic Books, 2000)

I have news for Professor Krauss. Both gravitating dark matter and the anti-gravitation dark energy accelerating the universe's expansion are simply negative and positive regions respectively of a single unified real Λ field. This Λ field is the residual random zero point energy fluctuation (zpf) contribution to the macro-quantum vacuum curvature term in Einstein's extended local geometrodynamic field equation for the bending of space-time by the stress-energy density of both real matter-radiation outside the vacuum and virtual "zero-point" matter-radiation inside the vacuum.

"the [zero-point] energy associated with the vacuum has precisely the form that results in a cosmological constant of the type Einstein invented ad hoc in 1916 ... We now have to ask not why the energy of empty space might be nonzero, but rather why it isn't much larger than is allowed by current observations... Somehow we have to explain how the cosmological constant could be at least 129 orders of magnitude smaller than we would naively estimate it should be. To date, no one has the slightest clue how this could result." Krauss, p. 104-5

Krauss's last sentence was true when written ~ 2000, but it is not true today 2002 because I have the explanation in this book.

Synopsis: The micro-quantum vacuum of special relativistic Schwinger source theory with zero expectation value $\left(8\pi G/c^4\right)\left\langle t_{\mu\nu}\left(zpf\right)\right\rangle = \Lambda g_{\mu\nu}$ is unstable against the formation of a macro-quantum Bose-Einstein condensation of a huge number of virtual electron-positron pairs at the fuzzy Heisenberg uncertainty edge of the Fermi-Dirac sphere into the same single particle bound state wave form for the phase-locked motion of the center of mass of each pair. The result is a holographic coherent phase order suppressing the random micro-quantum vacuum zero point fluctuations of all the quantum fields. This emergent "more is different" macro-quantum coherence gives Einstein's geometrodynamics with the new quintessent locally variable Λ field term.

Cal Tech's Fritz Zwicky at Mount Wilson in 1933 first inferred non-radiating gravitating dark matter in the fast relative motion within a group of galaxies ten million light years from us.

"There is now overwhelming evidence that more than 90% of the entire mass within the visible universe is made of material that is invisible to telescopes. The gravitational pull of this 'dark matter' therefore determines the motion of stars in galaxies, of galaxies in clusters of galaxies, and indeed of the universe itself." Krauss

The dark matter is not made of the same visible star stuff we are, i.e. protons, neutrons and electrons. The dark matter in and around the galaxies is ten times the visible galactic matter that is the same star stuff we are and it reaches ten times as far as the visible star stuff. Even that galactic dark matter is only one tenth of all the gravitating dark matter that is one hundred times more than all the star stuff we are.

I predict that the experiments to detect dark matter in the lab in the form of exotic particles will all fail! This is like the Michelson-Morley experiment to detect the motion of the Earth through the aether that also failed. If I am wrong about this, then my Λ theory is falsified in the sense of Karl Popper's criterion.

1. "the worst fine tuning problem in physics ... [is] the cosmological constant. If [it] is nonzero, but small, then a fine tuning of about 125 decimal places seems called for ... To date, no one even understands how to address the cosmological constant problem." p.142, Krauss

2. "Thus, whether dark matter or dark energy provides merely 90% of the mass in the universe today, as virial estimates indicate, or 99%, as the flatness argument suggests, there is now overwhelming reason to believe that all, or most dark matter is made from something else." p.168 Krauss

The Next Force

The Josephson Effect in Metric Engineering of Weightless Warp Drive
And
Star Gate Time Machines

Jack Sarfatti
Internet Science Education Project
San Francisco

Making Star Trek Real II

"Almost 100 years after Jules Verne another medium enriched the dream of spaceflight. The 3- year run of the initial TV series Star Trek, created by Gene Roddenberry, took us "where no man had gone before." Star Trek represented space flight in a way that became almost believable to the average person. Consequently, a whole generation was raised on the notion that Starfleet would make it possible for us to live and work in space in the not too distant future. This series became a true cultural phenomena with spin-offs of more TV series, established fan clubs, international conventions, and several major motion pictures. In later years, computer enhanced special effects allowed for the production of the Star Wars Trilogy, Close Encounters of the Third Kind, E.T., and Independence Day nourished the dreams of this generation and continue to build expectations of the promise of space travel.

> *"Where there is no vision, the*
> *people perish."*
> Proverbs 29:18

... Likewise, some future event will cause the dreams of the space cadets to be actualized into a new operational organization whose immediate focus is threat based, either from deep space or from a space faring nation in Earth's orbit.

... In all the space studies over the past 5 years, only three possible threat sources have been postulated: extraterrestrial aliens, hazards from asteroids and comets, and the WMD threat posed by rogue nations on Earth. Of these, the second is perhaps the most threatening to our existence as a nation and as a species.

> *"In our obsessions with antagonisms of the moment we often forget how much unites all the members of humanity. Perhaps we need some outside, universal threat to recognize this common bond. I occasionally think how quickly our differences would vanish if we were facing an alien threat*

103

> *from outside this world."*
> President Reagan,
> 21 September 1987

The "great communicator" made this remark[75] in context of a speech given to the UN General Assembly on the subject of arms control and world peace. This was the third public quote where the President wondered aloud about a threat external to our planet. If this ***significant emotional event*** were to emerge would we be forced to call on Astronauts like Bruce Willis or Robert Duvall? Hopefully we would be better prepared, but then all Titan II missiles have been dismantled. "
Col Victor P. Budura , Jr.
http://www.airpower.maxwell.af.mil/airchronicles/cc/budura.html
Thanks to Gary Bekkum for sending me this link.

The Equation of State for Zero Point Fluctuations of Gauge Fields

Quantum field theory is based on Einstein's theory of special relativity for a globally flat space-time.[76] For simplicity of explanation I only consider the quantum electrodynamics of electrons and photons. The spin ½ Dirac electron fermion spinor gauge source field has negative zero point energy density vacuum fluctuations $\rho_{zpe}c^2$ from a completely ionized neutral plasma of equal numbers of virtual off mass shell unbound polarized vacuum (PV) pairs of electrons and positrons. The spin 1 vector boson gauge force field has positive zero point energy density. Relativistic invariance requires that the zero point pressure p_{zpe} is always equal, though opposite in sign, to the zero point energy density for both fermions and bosons. That is, the zero point fluctuation equation of state for any quantum gauge source or force field is

$$\rho_{zpe}c^2 + p_{zpe} = 0 \tag{1.1}$$

The general relativity extension of Newton's gravity Poisson equation is Einstein's 1915 geometrodynamic field equation that can be written as

$$R_{\mu\nu} = -\frac{8\pi G}{c^4}\left(T_{\mu\nu} - \frac{1}{2}g_{\mu\nu}T_\lambda^\lambda\right) \tag{1.2}$$

(1.2) is equivalent to the usual form

$$G_{\mu\nu} \equiv R_{\mu\nu} - \frac{1}{2}R_\lambda^\lambda g_{\mu\nu} = -\frac{8\pi G}{c^4}T_{\mu\nu} \tag{1.3}$$

---

[75]The late United Artist Theaters motion picture billionaire who co-financed the James Bond films, Marshall Naify, and I had a significant role in this event. See Kim Burrafato's article in this book.

[76] Flat World , "Cosmological Physics" by John Peacock has these basic equations and GR sign conventions, e.g. pp. 25, 26. See also Peter Milonni's "Quantum Vacuum" for background information on positive and negative zero point energy densities for boson and fermion (respectively) special relativistic fields.

The stress-energy tensor for a relativistic fluid is

$$T_{\mu v} = \left(\rho + \frac{p}{c^2} \right) \frac{dx_\mu}{d\tau} \frac{dx_v}{d\tau} - p g_{\mu v} \tag{1.4}$$

Use (1.1) for the zero point fluctuations in (1.4). Therefore

$$T_{\mu v} \rightarrow t_{(zpe)\mu v} = -p_{zpe} g_{\mu v} = \rho_{zpe} c^2 g_{\mu v} \tag{1.5}$$

$$T_\lambda^\lambda = \rho c^2 - 3p \rightarrow t_{(zpe)\lambda}^\lambda = 4\rho_{zpe} c^2 \tag{1.6}$$

$$R_{\mu v} = -\frac{8\pi G}{c^4} \left(T_{\mu v} - \frac{1}{2} g_{\mu v} T_\lambda^\lambda \right) \rightarrow R_{(zpe)\mu v} = -\frac{8\pi G}{c^4} \left(t_{(zpe)\mu v} - \frac{1}{2} g_{\mu v} t_{(zpe)\lambda}^\lambda \right)$$

$$= -\frac{8\pi G}{c^4} g_{\mu v} \left(\rho_{zpe} c^2 - \frac{1}{2} 4\rho_{zpe} c^2 \right) = \frac{8\pi G \rho_{zpe}}{c^2} g_{\mu v} \tag{1.7}$$

$$\frac{G_{00}}{2} = R_{00} \rightarrow \nabla^2 \frac{U}{c^2} \tag{1.8}$$

$$T_{00} = \rho c^2 \tag{1.9}$$

$$T_\lambda^\lambda = \rho c^2 - 3p \tag{1.10}$$

$$R_{00} = -\frac{8\pi G}{c^4} \left(T_{00} - \frac{1}{2} g_{00} T_\lambda^\lambda \right)$$

$$= -\frac{8\pi G}{c^4} \left(\rho c^2 - \frac{1}{2} (\rho c^2 - 3p) \right) = -\frac{4\pi G}{c^4} (\rho c^2 + 3p) \tag{1.11}$$

$$\nabla^2 \left(\frac{U}{c^2} \right) = -\frac{4\pi G}{c^2} \left(\rho + \frac{3p}{c^2} \right) \tag{1.12}$$

Therefore, local classical field equation (1.2) in the weak curvature limit is Newton's Poisson equation with the relativistic pressure correction[77] Note the all-important factor of 3 in front of the pressure term. U is Newton's gravity potential energy per unit test mass with the dimensions of velocity squared. Hence the equation is one for *curvature* in the usual dimensions 1/Area. Therefore, in the special case of the quantum vacuum, substitute (1.1) into (1.12) to get

---

[77] Note that a positive $\rho + 3p/c^2$ gravitates, but a negative one anti -gravitates, e.g. "Foundations of Potential Theory", O.D. Kellog (1953) p. 156.

$$\nabla^2 U_{zpe} = 8\pi G \rho_{zpe} \tag{1.13}$$

Local Diff(4) covariance[78] plus the Einstein Equivalence Principle demands

$$t_{(zpe)\mu\nu} = \frac{\Lambda c^4}{8\pi G} g_{\mu\nu} \tag{1.14}$$

Where Einstein's local geometrodynamic field equation is now

$$G_{\mu\nu} + \Lambda g_{\mu\nu} = -\frac{G}{c^4} T_{\mu\nu} \tag{1.15}$$

Or, equivalently

$$G_{\mu\nu} = -\frac{8\pi G}{c^4}\left(T_{\mu\nu} + t_{(zpe)\mu\nu}\right) \tag{1.16}$$

Assuming zero torsion and metricity, the Bianchi identity for local momenergy[79] current conservation is

---

[78] That is the form of the local field equations remains invariant under arbitrary relative motions of local frames of reference that are momentarily coincident in a neighborhood of point space-time event P. Einstein's Equivalence Principle (EEP) is a stronger requirement on top of covariance, i.e., that there is a local tetrad transformation from a Locally Non-Inertial Frame (LNIF) at event P to a Locally Inertial Frame (LIF) at the same event P. Special relativity works in the LIF to a good approximation because the inhomogeneous curvature tidal forces of stretch-squeeze can be made small by making the space-time region of the LIF small. You need two test particles to measure the tensor tidal curvature force. You only need one test particle to measure the LNIF gravity pseudo-force that vanishes in an LIF like the Space Shuttle in orbit round the Earth with its rockets off. This single test particle pseudo-gravity force does depend on the local curvature in a static metric in the LNIF rest frame relative to the stress-energy source of that curvature. The pseudo-gravity force also depends on the distance of P to the source. The precise definition of this distance depends on the space-like symmetry of the source. The EEP approximation breaks down in the collapse of space-time to a singularity behind an event horizon of a black hole. The radii of curvature shrink to zero in the collapse to the singularity. "Same" and "momentarily coincident" mean a separation small compared to local radii of curvature. A big radius of curvature means weak curvature since the local curvature at event P associated with two directions is the reciprocal product of the two radii of curvature associated with those two directions. That is, curvature has the dimensions of 1/Area. Test objects in an LIF are weightless in free float. The test objects are on time-like geodesics. The time-like geodesic is the straightest path or "world line" in 4 dimensions that a material test particle can take in curved space-time. Time-like means inside the local light cone at P. The light cones at neighboring points tilt relative to each other in curved space-time leading to the Penrose diagram. A test particle is passive. It does not back-act on the geometry that pilots its motion. A test particle is not a source. This is a lesson that Bernie Haisch & Co have not yet learned. There is no such thing as "gravity force" as in Newton's theory. This is the meaning of "geometrodynamics", i.e. the elimination of force. Einstein considered "force" as part of the "wood" to be replaced with the "marble" of pure Platonic geometry. Objects in a LNIF are on time-like non-geodesics and they register weight on strain gauges from the non-gravity electrical reaction forces that pull the test objects off time-like geodesics. Given a bundle of neighboring time-like world lines that start from P_1 and end at P_2, the time-like geodesic has the longest frame invariant proper time between those points compared to time-like non-geodesics that intersect it at those points. This is the twin effect of differential aging depending on the path. This "critical point" in the functional space of world lines is a special case of the Action Principle of the calculus of variations out of which all the laws of physics local in space-time, or in its hyperspace generalization, come and must obey.

$$G_{\mu\nu}{}^{;\nu} = 0 \qquad (1.17)$$

This allows the transfer of momenergy current between the off mass shell random noise zero point fluctuations of virtual quanta and the on mass shell quanta as well as huge numbers of virtual quanta in macro-quantum Glauber coherent states[80] that are non-radiating external near fields. This is only for the gauge source and force quanta.

Vacuum Propeller Weightless Warp Drive

Note that for zero torsion

$$G_{\mu\nu}{}^{;\nu} \equiv \frac{\partial}{\partial x^{\nu}} G_{\mu\nu} + \Gamma_{\mu}^{\lambda\sigma} G_{\lambda\sigma} + \Gamma_{\nu}^{\nu\sigma} G_{\mu\sigma} \qquad (1.18)$$

However, if there is also torsion[81], but still metricity[82], then (1.17) is violated and the momenergy current density conservation equation is more generally

$$G_{\mu\nu}{}^{;\nu} + \frac{8\pi G}{c^4}\left(T_{\mu\nu}{}^{;\nu} + t_{(zpe)\mu\nu}{}^{;\nu}\right) = 0 \qquad (1.19)$$

The vacuum propeller warp drive equation is a piece of (1.19) when we can neglect the stress-energy density of both on mass shell gauge field matter and induction near fields to a good approximation, i.e.,

$$G_{\mu\nu}{}^{;\nu} + \Lambda^{;\nu} g_{\mu\nu} \approx 0 \qquad (1.20)$$

This is a direct transfer of momenergy current density from geometry to and from virtual zero point gauge quanta vacuum fluctuations bypassing the space-time stiffness barrier $G/c^4 \sim 10^{-33}$ cm per 10^{19} Gev.

---

[79] Components of the stress-energy tensor $T_{\mu\nu}$ of on mass shell real gauge source and force quanta as well as virtual gauge force quanta in macro-quantum coherent states forming non-radiating near induction fields.

[80] Possibly squeezed into a displaced ellipse in the Wigner density phase space of the field harmonic oscillator.

[81] Torsion $S_{\mu\nu}^{\lambda}$ is an antisymmetric 3rd rank tensor piece of the connection field $\Gamma_{\mu\nu}^{\lambda}$. Torsion does not vanish locally in the EEP tetrad map from an LNIF to a LIF. Torsion comes from topological dislocation gap defects in Hagen Kleinert's "world crystal lattice" model of GR. The lattice scale is the Planck scale $L_p = \sqrt{\hbar G/c^3}$. Curvature is from disclination topological lattice defects. These defects are singularities in the phase $\arg \Psi$ of the cohered virtual electron-positron bound state Bose-Einstein condensate.

[82] Metricity is vanishing covariant divergence of the geometrodynamic field $g_{\mu\nu}$. This can be destroyed from extra dimensions of hyperspace.

IT FROM BIT? Wheeler's Self-Organizing Universe.

We shall see that Λ is strongly coupled to the gauge force fields via the macroscopic vacuum coherence order parameter $\Psi(x)$ that is a complex scalar field in curved space-time. Yet, its phase $\arg\Psi$ determines curved space-time $g_{\mu\nu}$ in a self-organizing feedback loop between geometry (Wheeler's "IT") and Bohm giant pilot wave (Wheeler's "BIT"). That is John Archibald Wheeler's "self-excited universe" is "IT FROM BIT" + "BIT FROM IT". Wheeler was missing the "BIT FROM IT" part of the equation.

The Two Faces of Janus - Dark Energy and Dark Matter

Returning to (1.4), in general for the relativistic fluid

$$T_{\mu\nu} = \left(\rho + \frac{p}{c^2}\right)\frac{dx_\mu}{d\tau}\frac{dx_\nu}{d\tau} - pg_{\mu\nu} \tag{1.21}$$

Therefore, for the local random zero point fluctuations in particular from (1.1)

$$t_{(zpe)\mu\nu} = -p_{zpe}g_{\mu\nu} = \rho c^2 g_{\mu\nu} \tag{1.22}$$

The metric signature with the present sign conventions is +—-. Therefore

$$t_{(zpe)00} = \rho c^2 \equiv \frac{\Lambda c^4}{8\pi G} \tag{1.23}$$

Consequently (1.13) is

$$\nabla^2 U_{zpe} = 8\pi G\rho_{zpe} = c^2\Lambda \tag{1.24}$$

Where $\Lambda > 0$ is a local universally repulsive anti-gravitating "dark energy" vacuum region of positive zero point fluctuation virtual boson energy density. Furthermore, $\Lambda < 0$ is a local universally attracting gravitating "dark matter" polarized vacuum region of negative zero point fluctuation energy density made out of virtual fermion-antifermion pairs that are ionized not in a bound state.

Unconventional Flying Objects

Think of the zero point Λ field as a new kind of *virtual* spin 2 charge. Two opposite Λ charges rigidly connected to each other will self-accelerate. The energy comes from the vacuum. This is the Bondi vacuum propeller[83] effect. This is very different from opposite spin 1 electrical charges that attract and like electrical charges that mutually repel conserving the center of mass motion. Imagine a flat high T_c superconducting circular plate with $\Lambda > 0$ on bottom and $\Lambda < 0$ on top. The plate will accelerate vertically upward. Strain gauges imbedded in the plate will show no g-force. The conjectured electromagnetically controlled local Λ field configuration generates its own free float time-like geodesic warp drive. You can get it to hover or do anything you like in principle. Sounds like a flying saucer, eh? – like Paul Hill's[84] "acceleration field"? You got it Harry Potter! No engines, everything is done at the micro-nano engineered level inside the skin of the fuselage like Colonel Phillip J. Corso reported.

More Is Different: Vacuum Instability

The Dirac electron *micro-quantum false vacuum* is a filled Fermi sphere of closely packed negative energy electrons in globally flat 3-momentum space complementary to globally flat 3D space. *There is no gravity at all in this Flat World.* The momentum at the surface of the Fermi sphere is of order $\hbar/L_p \sim 3\times10^{-5}\times10^{10}\, gm\times cm/\sec$. The Heisenberg uncertainty principle fuzzes out the sharp edge of this Fermi sphere over a region of order $2mc \sim 6\times10^{-27}\times10^{10}\, gm\times cm/\sec$. The ratio of fuzzy thickness of the surface to the radius of the Fermi sphere is $\sim 10^{-22}$. Therefore, as in BCS superconductor theory, the binding energy of a single virtual electron-positron (hole) pair in the Heisenberg fuzzy edge under their mutual Coulomb attraction of unlike electrical spin 1 charges, is of order $\alpha\, m_p c^2 \sim 10^{17}$Gev. The photon rest mass is 10-65 gm with Meissner penetration depth 10^{28} cm. Therefore this is a hard superconductor with a huge varying penetration to coherence ration and quantized vortex string singularities in the phase field. The Bose-Einstein condensation energy density for the creation of the universe is $\sim 10^{17}$Gev L_p^{-3} (m/mp) $\sim 10^{26}\, 10^{99}\, 10^{-22}$ ev/cc. That's big Bhubba! However, the pair is off mass shell. Therefore, the Flat World *micro-quantum* false vacuum[85] is unstable to the formation of the Curve World paired *macro-quantum* vacuum. That is, this *single pair* bound state wavepacket $\psi(x)$ is macroscopically occupied by virtual electron-positron pairs with a superfluid number density $|\Psi(x)|^2$. There is a negative effective Bose-Einstein condensation energy density in the "More is different"[86] phase transition from the micro-quantum false flat world vacuum of completely ionized virtual

---

[83] The term "vacuum propeller" was coined by Roger Coolidge's laboratory assistant Igor Kelvin at the secret laboratory in the mountains of the Duchy of Grand Fenwick.☺ See "Tuxedo Park" by Jennett Conant and "Gravity's Rainbow" by my Cornell Classmate Tom Pynchon. I attended lectures at Cornell in ~1960 by Herman Bondi, Chief Scientist of the British Ministry of Defense on this self-acceleration idea with "negative mass" 42 years before I discovered its connection to the Λ field.

[84] "Unconventional Flying Objects"

[85] The vacuum of Haisch-Rueda-Puthoff in SED ZPE origin of inertia model is this false Flat World vacuum which gives the wrong Λ too big and positive (anti-gravitating) by 122 powers of ten. Similarly for Puthoff's PV gravity model which is, in addition, wrong for other reasons such as the infinite set of isotropic radial coordinates he needs going complex when the unique curvature coordinate is inside the gravity radius. Puthoff's "PV without PV" suffers from excess mathematical baggage and ill-posed fragmentary mutually inconsistent conceptualizations of the physical picture of reality.

[86] P. W. Anderson, "A Career in Theoretical Physics" (World Scientific)

electron-positron plasma at the fuzzy edge of the Fermi sphere surface to the macro-quantum true curve world vacuum consisting of a huge number of bound virtual positronium pairs all in the same wave packet $\psi(x)$ in their *center of mass* coordinates x.[87] Thus, the metastable macro-quantum curve world vacuum is lower in energy density relative to the unstable micro-quantum flat world false vacuum by an amount ~ $|\Psi(x)|^2 2m_e c^2 \sim 10^{99} \times 10^{-3} \, Gev/cm^3$ for the Einstein $\Lambda = 0$ nongravitating phase of vacuum that we are used to here on Earth under ordinary physical conditions. This smooth coherent "signal" superfluid curve world vacuum has zero thermodynamic entropy. This explains the zero entropy initial condition on the universe that leads to the Arrow of Time in which we, in our consciousness, experience a forward flow of subjective time in which we objectively grow old and die as the universe expands keeping the sky dark and space cold. You can think of the macro-quantum vacuum $\Psi(x)$ as a generally locally covariant spin 0 scalar zero rank tensor field in the complex plane. Einstein's local geometrodynamic curve world metric field is simply, in the absence of external gauge force fields

$$g_{\mu\nu}(x) = \eta_{\mu\nu} + \frac{L_p^2}{2}\left(\frac{\partial}{\partial x^\mu}\frac{\partial}{\partial x^\nu} + \frac{\partial}{\partial x^\nu}\frac{\partial}{\partial x^\mu}\right)\arg\Psi(x) \qquad (1.25)$$

Where $\eta_{\mu\nu}$ is the globally flat space-time Minkowski metric. The phase dependent second term in RHS of (1.25) need not be small. This is not a linearized theory. The quintessent locally variable $\Lambda(x)$ field is simply

$$\Lambda(x) = \frac{1}{L_p^2}\left(1 - L_p^3 |\Psi(x)|^2\right) \qquad (1.26)$$

Wavelet Scale Dependence & Wigner Densities

Everything is scale-dependent in the sense of adaptive window wavelet transforms replacing the rigid Fourier transform that is only good in global flat world. I suppress that detail here for simplicity of pedagogical exposition. The Wigner phase space density[88] is replaced by a Wigner scale space density[89] in which a wavelet kernel replaces the Fourier kernel $e^{ip_\mu x^\mu}$.[90]

---

[87] Note that the virtual electron and virtual positron 3-current densities add constructively under an applied electric field. Indeed, these PV 3-currents may be Maxwell's vacuum displacement 3-current from the rate of change of the mixed space-time components of the dual Hodge star electromagnetic Cartan 2-form *F (field tensor in engineering language). That is, Maxwell's equations are topologically dF = 0 and *d*F = J. The virtual electron-positron pairs are in the * part of the field equations. This is an example of Wheeler's "the boundary of a boundary is zero".

[88] "Phase Space Picture of Quantum Mechanics", Kim & Noz (World Scientific)

[89] The Wigner density is real but unlike a classical probability density it can go negative from quantum superposition. This happens on a cosmic scale in the dark matter!

[90] p. 63 eq. (3.6) "A Friendly Guide to Wavelets", Gerald Kaiser, Birkhauser

$$\psi_{s,x}\left(u\right)=\left|s\right|^{-\eta}\psi\left(\frac{u-x}{s}\right) \tag{1.27}$$

The scale is s, and you can think of ψ as the bound state wave packet of a single virtual electron-positron pair. The residual Heisenberg uncertainty random noise field has a Wigner scale space density

$$\rho_n\left(x,s\right)\equiv\frac{\Lambda\left(x,s\right)}{L_p} \tag{1.28}$$

Just as in the two-fluid theory of Helium II, the thermodynamic entropy density is carried only by this residual random micro-quantum noise field. This entropy density of the macro-quantum vacuum may be, *I conjecture* something like

$$S\overset{?}{=}kL_p^3\rho_n\ln\left(1+L_p^3\rho_n\right)=kL_p^2\Lambda\ln\left(1+\left|L_p^2\Lambda\right|\right) \tag{1.29}$$

Note, that if equation (1.29) is true, then gravitating $\Lambda<0$ dark matter has *complexity-generating negative entropy* consistent with the clumping needed to form galaxies and stars.

The Josephson Effect in Metric Engineering

Now imagine a weak Josephson link[91] between the Bondi capacitor real electron pair superconductor and the vacuum virtual positronium superconductor. P W Anderson points out that the micro-quantum dynamics, in this case virtual electron-positron pairs binding from their Coulomb attraction inside the vacuum on the one hand, real electron pairs bound by phonons on the other hand, are not important to the emergent collective "More is different" macro-quantum order parameter Ψ acquiring a life of its own on the larger scale. Therefore, we do not think of a literal propagation of either virtual electron-positron pairs or real electron positron pairs through the conjectured weak link between vacuum and superconductor, rather we think of a nonlocal reach of the macro-quantum version of Bohm's quantum potential Q to ensure conservation of momenergy currents between vacuum and superconductor. This means a nonlocally correlated mutually compensating adjustment in the $\left|\Psi\right|^2$ on both sides of the weak link. You need virtual positronium vacuum superconductor, *not* "dilithium crystals", to Make Star Trek Real and to make the curved space-time without which nothing material is possible. The effective quintessent field is then

---

[91] Josephson Effect, Vol III Feynman's Lectures on Physics, Ch. 21. Oddly Dennis Schmidt in "A Satori Trilogy" written maybe 20 years ago writes of a Sarfatti-Josephson Star Drive. Kim Burrafato witnessed this strange incident finding the Schmidt story in a bookstore in the SF Marina. He may have read about Brian visiting me in 1976 that was in the San Francisco Chronicle. But my idea was still 25 years in the future. I have not communicated with Brian Josephson for 5 years now.

Jack Sarfatti

$$\Lambda = \frac{1}{L_p^2}\left(1 - L_p^3 \left|\Psi_{vac} + \Psi_{sc}\right|^2\right)$$

$$\left|\Psi_{vac}\right| >> \left|\Psi_{sc}\right|$$

$$\therefore$$

(2.1)

$$\Lambda = \frac{1}{L_p^2}\left(1 - L_p^3 \left|\Psi_{vac}\right|^2 \left|1 + \frac{\Psi_{sc}}{\Psi_{vac}}\right|^2\right)$$

$$\approx -2L_p \left|\Psi_{vac}\Psi_{sc}\right|\cos\left(\arg\Psi_{vac} - \arg\Psi_{sc} - \frac{2e}{\hbar c}\oint A_\mu dx^\mu\right)$$

Plugging in the numbers gives

$$\Lambda \approx -2\times10^{-33}\times10^{45.5}\times10^{11.5}\times\cos\Theta$$

$$\rightarrow 10^{24}\cos\Theta\, cm^{-2}$$

(2.2)

Where the gauge invariant Josephson weak link relative phase difference connecting vacuum geometrodynamics to the real superconductor disk is

$$\Theta \equiv \arg\Psi_{vac} - \arg\Psi_{sc} - \frac{2e}{\hbar c}\oint A_\mu dx^\mu$$

(2.3)

There will also be a Chiao "gravity radio" gravimagnetic-magnetic coupling in the space-like 3-vector part of the phase. Note that the cosine function has a nonlinear Taylor power series expansion. Apply an AC voltage at resonance to get a DC momenergy current transfer between vacuum and superconducting matter. There are several interesting effects here to be worked out later.

Note that the effective scale of variable Λ field radius of curvature is

$$r_\Lambda \approx \frac{10^{-12}}{\sqrt{\left|\cos\Theta\right|}}\, cm$$

(2.4)

This is enormously strong when the cosine is near 1. The corresponding curvature radius from the mass of the Earth at the surface of the Earth is about 1 AU $\sim 10^{13}$ cm. Therefore, the Λ field is effectively at least 25 powers of ten stronger in warp power than gravity at the surface of the Earth. The aerospace weapons potential of my discovery here dwarfs that of nuclear fusion and may explain the real physics of ultra-violent astrophysical events including gamma bursters.[92] One peaceful use of my Λ field "metric engineering"[93] "Making Star Trek Real"[94] will be to divert huge asteroids on a collision course with Earth.

---

[92] The large extra space dimensions brane idea of Dvali et-al in "Scales of Gravity" with a larger scale quantum gravity cutoff $L_p^* >> L_p$ make this effect weaker by a factor $(L_p^*/L_p)^{1/2}$ multiplying 10^{-12} in (1.33).
[93] Coined by Hal Puthoff.
[94] Coined by me.

Gravity Radio C³?

Ray Chiao[95], distinguished professor of physics at UC Berkeley, has an intriguing idea for the efficient practical conversion of gravity waves to electromagnetic waves and vice versa inside Type II superconductors[96] that support vortices of quantized magnetic flux

$$\Phi_0 \equiv \frac{h}{2e} = 10^{-7} Gauss \times cm^2 \tag{3.1}$$

rather than the complete Meissner effect expulsion of magnetic field in an ideal Type I superconductor. Chiao needs to impedance match the lowest order quadrupole gravity wave to a quadrupole electromagnetic wave.

This would be perfect for the underwater nuclear submarine and aerospace forces since gravity waves cannot be stopped by any barrier at least when $\Lambda = 0$.

Far field radiation is generally weak compared to near induction fields. That's a good thing too. "God is subtle but not malicious."[97] As metric engineers we are interested in the stronger non-radiating near induction fields. Radiation is bad for virtual vacuum state metric engineering of warp drives and Star Gates because it leaks away energy like a hole in a water pipe or a badly dripping faucet. Einstein always said he would be a plumber if he had to do it all over again. Indeed, Lenny Susskind was a plumber before Peter Carruthers got him into Cornell in 1963 when I first met him and turned him on to the time-phase quantum operator problem.

Look at the Josephson phase in (2.3). It is both gauge invariant and space-time invariant. Ray Chiao gravimagnetic-magnetic coupling is non-relativistic but gauge invariant in the sense of minimal coupling. It is not however 4-dimensionally relativistic invariant. Therefore I add an extra term to what Ray Chiao did in order to use "gravity warp induction" in the Josephson weak link between the real superconducting high T_c Bondi vacuum propeller and the macro-quantum vacuum. Chiao's interaction Hamiltonian density is

$$H_{int} = \left|\Psi(x)\right|^2 2e\vec{A} \cdot \vec{H} = \left|\Psi(x)\right|^2 2eA^i g_{0i}$$
$$i = 1,2,3 \tag{3.2}$$

I change this to

$$H_{int} = \left|\Psi(x)\right|^2 2eA^\mu H_\mu = \left|\Psi(x)\right|^2 2e\left[A^i g_{0i} - A^0 \left(g_{00} - \eta_{00}\right)\right] \tag{3.3}$$

---

[95] "Superconductors as transducers and antennas for gravitational and electromagnetic radiation"
http://xxx.lanl.gov/abs/gr-qc/0204012

[96] There are two length scales in a superconductor, the coherence length ξ over which Ψ varies and the London penetration depth λ over which the magnetic flux varies in the Meissner effect. This is an exponential decay in the case of a real superconductor. Type II has $\xi < \lambda$. The vacuum will have these as variable local fields themselves scale-dependent in the sense of wavelet transforms.

[97] Albert Einstein

Note, that in the static weak field limit in the quasi-uniform gravity field at the surface of the Earth.

$$H_{int} \rightarrow -\left|\Psi(x)\right|^2 2eA^0 \frac{gz}{c^2} \tag{3.4}$$

I may be off by a minus sign. ☺ Of course A^0 is an electrical voltage. Therefore, I predict on the basis of making Chiao's gravity radio locally covariant, consistent with curved space-time, not the Haisch-Puthoff-Firmage "Flat World", a direct superconducting amplified voltage-gravity coupling that avoids the G/c^4 space-time stiffness barrier.[98] This is not directly the same as my Josephson effect however, but it is related to it. Let's go back to that.

$$\Theta \rightarrow \arg\Psi_{vac} - \arg\Psi_{sc} - \frac{2e}{\hbar c}\oint A_\mu dx^\mu - \frac{mc}{\hbar}\oint H_\mu dx^\mu \tag{3.5}$$

with nonlinear gravity-electromagnetic couplings inside the cosine that is a Taylor power series expansion.

A note on the meaning of "virtual off mass shell" and "real on mass shell" in perturbative quantum field theory.

Given a spin 0 massive[99] scalar boson quantum field ϕ for simplicity. We can use the Fourier transform in flat space-time. We need to use wavelet transforms in curved space-time. Forget that complication for now. The perturbative response of the field $\phi(x)$ at field point P with coordinates x to a an arbitrary source distribution $J(x)$ is the 4-D Fourier integral

$$\delta\phi(x) = \int d^3k e^{i\vec{k}\cdot\vec{x}} \int d\omega e^{-i\omega t} \frac{J(\vec{k},\omega)}{\vec{k}^2 + \left(\frac{mc}{\hbar}\right)^2 - \left(\frac{\omega}{c}\right)^2} \tag{4.1}$$

The Green's function propagator in 4-momentum space is

$$\tilde{G}(\vec{k},\omega) = \frac{1}{\vec{k}^2 + \left(\frac{mc}{\hbar}\right)^2 - \left(\frac{\omega}{c}\right)^2} \tag{4.2}$$

The mass shell is the pole of this propagator, i.e. where the denominator vanishes

---

[98] G is still hidden in g of course, but this new Modanese type macro-quantum amplified direct gravity-gauge force coupling is not a direct stress-energy bending of space-time that is limited by the barrier.
[99] The rest mass of a real quantum of the field on mass shell is m. If m is imaginary we have a real tachyon.

$$\vec{k}^2 + \left(\frac{mc}{\hbar}\right)^2 - \left(\frac{\omega}{c}\right)^2 = 0$$

$$\omega_{\pm} = \pm c \sqrt{\vec{k}^2 + \left(\frac{mc}{\hbar}\right)^2}$$

(4.3)

The integral (4.1) has two parts: the "real on mass shell" part from the pole and the "virtual off mass shell" part from everything else. In the case of the Maxwell spin 1 four vector[100] potential field A_μ, the pole is the Fourier transform of the light cone and its contribution is far field transverse polarized radiation leaking energy to infinity. The virtual off mass shell non-pole part are virtual photons of all three independent polarizations including the longitudinal polarization in the direction of the 3-momentum vector $\vec{k}$. The two transverse polarizations for each $\vec{k}$ mode (quantum harmonic field oscillator) are in the plane perpendicular to $\vec{k}$. This plane wave decomposition into modes is not good inside waveguides and cavities like in the old microwave klystron et-al. The fourth time-like polarization is not independent of the longitudinal polarization. They are connected by the Lorentz gauge constraint since a quantum spin 1 field must have only 3 independent spin polarization projections relative to an arbitrary space direction. Since the photon has zero rest mass, the inductive longitudinal part of the field cannot leak energy to infinity in the form of radiation. The photon gets some rest mass inside a superconductor. This is part of the Meissner effect.

Look at the time-frequency integral in (4.1)

$$\int d\omega e^{-i\omega t} \frac{J\left(\vec{k}, \omega\right)}{\vec{k}^2 + \left(\frac{mc}{\hbar}\right)^2 - \left(\frac{\omega}{c}\right)^2}$$

(4.4)

This integral is not well defined without boundary conditions that are closed paths or contours in the complex frequency plane $\omega = \omega_r + i\omega_i, i^2 = -1$, or complex energy plane since $E = \hbar\omega$. There are now choices. The contour used is Feynman's. Since there are actually two complex ω poles or zeros of the denominator (1.41) on the real ω axis on

---

[100] This is a first rank tensor equivalent to a second rank Penrose spinor of the 10 parameter Poincare Lie group for the globally flat space-time of special relativity. When the 4-parameter translation group generating total energy and total linear momentum conservation (Emmy Noether's theorem) is locally-gauged away, like in non-Abelian Yang-Mills internal gauge symmetry fields of the electro-weak-strong force, we get the full nonlinear curved space-time of Einstein's general relativity. This was first done by Kibble at Imperial College, London during Abdus Salam's era ~ 1966. I was there as a visitor down from UKAERE, Harwell. The tensors and spinors are then those of the Diff(4) local group. The idea of Bernie Haisch and Hal Puthoff, promoted by Joe Firmage, that curved space-time is not a good idea is pure Cargo Cult pseudo-physics in the sense of Richard Feynman's famous Cal Tech Lecture that warned of the dangers of New Age psychobabble coming from est and Esalen in Big Sur, California. He visited there because John Lilly was his friend and Esalen was full of friendly scantily dressed pretty women.

equally distant opposite sides of the imaginary ω axis[101], Feynman circles around the two poles with two opposite infinitesimal half circles and closes the contour with a big half circle of radius $\to \infty$ in such a way that, for the spin 1 photon field with $m \to 0$, positive real energy is a far field transverse radiation retarded wave along the forward in time future light cone, and negative real energy is an advanced wave along the backward in time past light cone. You actually need advanced waves from the future to get the correct very accurate answers in the perturbation expansion of Feynman diagrams made out the Green's function propagators for quantum electrodynamic phenomena like the Lamb shift, anomalous magnetic moment of the electron et-al. Indeed, for all $m \neq 0$ quantum fields of any spin, real on mass shell retarded antiparticles of positive energy forward in time are *equivalent to* advanced particles of negative energy and opposite charges backward in time using the particular Feynman contour in the complex energy plane of the integral (4.1).

More Is Different

The actual equations here are for a massive spin 0 scalar boson field. The basic idea works for all spins including the spin ½ Dirac electron, the spin 1 vector gauge bosons of the electro-weak-strong forces of nuclear physics, the spin 3/2 field and the spin 2 graviton field. The spin 2-graviton field is a concept of limited usefulness. Penrose[102] calls it the "linear graviton" as distinct from his "nonlinear graviton" connected with the non-perturbative background independent attempts at quantum gravity involving twistors and instantons in imaginary time, spin networks et-al. The key point for us at the moment, is that you cannot get Curve World from Flat World by taking a finite number of spin 2 linear gravitons propagating in Flat World anymore than you can get a superconducting macro-quantum Bose-Einstein condensate of paired electrons[103] from the micro-quantum Dirac-Fermi Heisenberg fuzzy sphere densely packed with negative energy virtual electrons[104] with a finite number of attractive electron-phonon perturbation Feynman diagrams. In both cases you need to sum an infinite series of Feynman diagrams. This is beyond perturbation theory. This is P.W. Anderson's "More is different" of emergent collective order in action. The formation of complex macro-quantum local order parameters $\Psi(x)$ with long-range phase coherent hologram interference patterns is essentially non-perturbative. You can think of micro-quantum random gravitons as normal fluid fluctuations from a macro-quantum

---

[101] As the rest mass control parameter $m \to 0$, the two distinct frequency roots $\omega_{\pm} \to \pm c|\vec{k}|$ correspond to advanced and retarded electromagnetic waves for the spin 1 Maxwell field with gauge invariant minimal coupling to the electron-positron 4-component Dirac spinor field and the longitudinal spin-polarization does not propagate as a far field.

[102] "The Geometric Universe" (Oxford, 1998), ed. Huggett et-al. " the … twistor programme in which sheaf cohomology groups … corresponded precisely to the solutions of the zero rest mass field equations … a complexification of the Radon transform … for application to tomography … twistors …solve the self-dual Yang-Mills equations … this stimulated work on instantons and … led to Donaldson's … work on 4-manifolds … and … led to a deep understanding of the self-dual Einstein equations in which the Riemannian geometry gets encoded entirely into the holomorphic geometry of a complex 3-manifold…hyper-Kahler manifolds are the natural generalizations of self-dual Einstein manifolds … which …arise … in supersymmetric gauge theories." M. Atiyah

[103] Or paired virtual electrons and positrons exchanging attractive space-like virtual photons in the real PV vacuum case not Hal Puthoff's Mickey Mouse Cargo Cult version.

[104] One per Fermi field oscillator mode required by Pauli exclusion antisymmetry permutation principle.

graviton superfluid Bose-Einstein condensate that is curved space-time. However, that graviton condensate is itself a kind of Goldstone collective boson coherent phase mode of the virtual electron-positron vacuum condensate.

"if we remove life from Einstein's beautiful theory by steam-rolling it first to flatness and linearity, then we shall learn nothing from attempting to wave the magic wand of quantum theory over the resulting corpse." Roger Penrose [105]

Harnessing The Cosmic Energy of The World Hologram

——-Original Message——-
From: Black Ops Agent X
Sent: Friday, September 06, 2002 7:59 AM
To: sarfatti@well.com
Subject: Next Force Question

"Jack, In Next Force after equation (1.24) you explain: "Where $\Lambda > 0$ *is a*
locally universally repulsive antigravity "dark energy" vacuum region of positive zero point fluctuation virtual boson energy density. $\Lambda < 0$ *is a locally universally attracting gravitating "dark matter" polarized vacuum region of negative zero point fluctuation energy density made out of virtual fermion-antifermion pairs that are ionized not in a bound state."* *Are you saying that the superconducting rotating disks of the Bondi capacitor or vacuum propeller are to be designed and engineered so that the antigravity "dark energy" portion of* VACPROP *engages near fields of virtual photons, thus* $\Lambda > 0$, *and the gravitating "dark matter" portion engages the ionized virtual positronium BEC in a macro quantum superfluid state, thus* $\Lambda < 0$?

The result being that upon engagement, the dark matter portion attracts and the dark energy portion repels causing VACPROP *to self propel?"*

Jack: No. Do not think in terms of something engaging something else. It just is. The dark energy has a preponderance of virtual photons compared to virtual electron-positron plasma. The dark matter is the opposite. It's a matter of tipping the balance one way or the other. One balance dish is virtual bosons. The other balance dish is virtual electron-positron plasma. Where you put the pivot point of Archimedes Lever is the macro-quantum coherent order parameter $\Psi(x)$. The point is that the deep structure of physical vacuum in my theory is variable and controllable in principle.

It is standard micro-quantum field physics that the random virtual zero point bosons naturally universally anti-gravitate repulsively and that the random virtual unbound fermion-antifermion plasma universally gravitate. The physical micro-quantum false vacuum is this alphabet soup of randomly fluctuating virtual bosons and virtual fermion-antifermion pairs. That's what micro-quantum vacuum is!

The Λ local field is the net effect of all of these random zero point vacuum fluctuations. We see the possibility that $\Lambda = 0$ since bosons of spin 0, 1 and fermions of spin ½ & 3/2

---

[105] In Abhay Ashtekar's paper in "Geometric Universe" that shows why the Haisch-Puthoff-Firmage Flat World theory is simpler than is possible.

pull Λ in opposite directions. Hence $\Lambda = 0$ is plausible. $\Lambda = 0$ is what we normally think of the vacuum as gravitationally neutral.

Note I left out spin 2 gravitons for a deliberate reason! They come in much later after the vacuum phase transition from Flat World to Curve World that spontaneously breaks symmetry. Which symmetry? Why Poincare symmetry of course! I mean the translational symmetry subgroup of the Poincare group that is locally gauged away! What does that mean? That means that space-time is no longer a rigid playing field that pilots or grips matter, causing its motion, without any direct back-action on it from matter.[106] Space-time is now an active participant a source and sink of energy and linear momentum via the space-time stiffness valve G/c^4. This stiffness valve is very asymmetric. Gravity from a huge mass has a much stronger effect on matter in motion than matter in motion has on gravity. You can see that when you climb the stairs or fall off a ladder. In comparison it takes an enormous amount of electrical energy to directly deform the gravity tidal force curvature field! Indeed it takes 4 billion tons of mass energy equivalent to make a 1-fermi change in the local radius of curvature! That's space-time stiffness Bhubba!

What is real? The Looking Glass MIND WARS

Fortunately for the Army, the Navy and the Marines, I have found a way to leap frog over this space-time stiffness barrier with the Λ-field. I have found a way, a Tao of Thought as it were, to quantitatively begin to do the R&D needed to obey General Douglas MacArthur's final order to the cadets at West Point "Harness the cosmic energy" in "Duty, Honor, Country". The Λ field is precisely the "cosmic energy" our eloquent General was alluding to. The UFO writer Colonel Phillip J. Corso was on MacArthur's Staff and rumor is that MacArthur and CIA's James Jesus Angleton were both alarmed about invasion from flying saucers. I have no idea if there is any truth to that of course, but those stories are "Out There" in Cyber Space. The case of Edward Teller is enigmatic. Stephen Schwartz, who has worked with Robert Conquest at the Hoover Institution[107] and with Donald Rumsfeld[108] at the Institute for Contemporary Studies, America's leading USG expert on Soviet espionage in America with access to the Venona File, reports that Teller said that the UFO craze in this country was started years ago by the Soviet KGB "to weaken American Science" using phony Majestic 12 papers[109] and New Age organizations like est and Esalen. Also I heard Teller debunk UFOs at a fundraiser for the Hebrew Academy in San Francisco and I spent

---

[106] Source matter, including all electromagnetic fields, by definition, back-acts on space-time geometry via the stress energy density tensor $T_{\mu\nu}$ channeled through the tiny G/c^4 space-time stiffness reciprocal super string tension valve transforming stress-energy density into curvature by the direct brute force process Einstein discovered in 1915 as a generalization of Newton's 17th Century gravity theory. In contrast, a test particle with nonzero real rest mass, by definition, does not back-act directly on the space-time that is piloting its motion along a time-like geodesic along the straightest paths in 4D-curved space-time. The flying saucer, *I conjecture*, is a new breed that shapes its own time-like geodesic path manipulating the Λ field with on board electromagnetic nonradiating induction near field generators made out of high Tc superconducting smart nano-materials.

[107] See paperback "Special Tasks" by Sudoplatov with the forward by Robert Conquest.

[108] Secretary of Defense in the second Bush Administration.

[109] Joe Firmage supported new publicity on MJ 12 in 1999 in the work of Dr. Robert Wood that Stephen Schwartz says is the same old rehashed Soviet KGB disinformation. Another former high CIA official confirmed the phoniness of MJ 12 to me by email in September 2002.

several hours with Teller and Itzak Rabin at the home of Rabbi Pincus Lipner. On the other hand, in a recent 2002 meeting I had in North Beach with a high-ranking CIA officer I will call "Deep Throat", I was told that indeed Teller was quite concerned with a UFO threat in the early 80's when he was lobbying the Reagan White House for the Strategic Defense Initiative that I also, with Marshall Naify, played a role in.

Flat World is an unstable false vacuum!

This micro-quantum vacuum lives in globally flat space-time, i.e. the Flat World of Haisch-Puthoff-Firmage Yilmaz theory. But there is no gravity as yet! This is the Andre Sakharov idea of 1967 that gravity is an emergent collective order, a "lumped parameter" like inductance and capacitance, or like temperature in kinetic theory of gases etc, out of the gauge source and force global special relativity quantum fields.

This Flat World is Gauge World.

According to Dirac, the spinor electron part of the micro-quantum vacuum is a filled Fermi sphere of negative energy electrons one per mode in phase space in accord with the Pauli exclusion principle of complete finite permutation group antisymmetry of the thought like pilot quantum information wave of several identical fermions. The radius of this Fermi sphere in the momentum subspace of phase space is ~ h/(Planck length) since the Planck length 10^{-33} cm is the smallest length because of Heisenberg's uncertainty principle. Heisenberg's principle also demands that the Fermi sphere not be classically sharp like a marble. It has a fuzzy edge of thickness ~ twice the Compton momentum of the electron. This thin spherical fuzzy shell is where you find the virtual electron-positron completely ionized plasma isre called the polarized vacuum (PV) zero point fluctuations. Of course Hal Puthoff's theory does not have this. There is no PV in his PV theory!

But Flat World is an unstable false vacuum!

Opposite electric charges attract. Therefore, the virtual electron-positron pairs in this thin Heisenberg uncertain fuzzy edge of the Fermi sphere form a macro-quantum Bose-Einstein condensate with a local giant thought like pilot wave of information $\Psi(x)$ exactly like in the BCS picture of superconductivity. In the latter, real electron pairs form because of an attractive sound wave (phonon) coupling. We don't need that here.

This is the **"More is different"** vacuum phase transition from Flat World to Curve World in which Einstein's local geometrodynamic field is simply a "world hologram" from the phase interference of the thought like giant Ψ field. The residual random zero point quintessent Λ field is a simple function of the intensity of the world hologram!

Curve World = World Hologram[110]

---

[110] Lenny Susskind means something slightly different by the term "world hologram". However the two meanings are connected. Each Planck area L_p^2 is a c-bit of information in the thermodynamics of black holes. Given a 3D volume of space, Lenny's conjecture is that the maximum information that can be packed in that space is ~ $V^{2/3}/L_p^2$. This means our visible universe has perhaps roughly (Hubble Radius)^2/(Planck Area) = 10^{122} c-bits of Shannon information. This increase of information/entropy as the space of the universe expands is related to The Arrow of Time. More precisely take $R(t)^2/L_p^2$ in the FRW cosmic metric with cosmic time dependent scale factor R(t) in length units.

Now to answer your question of how to make our own flying saucer to defend ourselves against any possible alien ET threat, or even from an asteroid on a collision course with Earth, $\Lambda > 0$ phase of vacuum is the universally repulsive anti-gravity phase which happens when the quintessent local intensity of the world hologram is too weak relative to the critical value that makes $\Lambda = 0$. Obviously then, $\Lambda < 0$, the universally attractive phase happens when the local intensity of the world hologram is too strong. Since virtual random zero point bosons make $\Lambda > 0$ and virtual random (ionized plasma) zero point fermion-antifermion pairs make $\Lambda < 0$, I call the weak intensity region of the hologram "boson dominated" and the strong intensity regions "fermion dominated".

Think of the world hologram as a landscape of hills and valleys. The valleys are $\Lambda < 0$ dark matter, the hills are $\Lambda > 0$ dark energy. The actual pattern of the world hologram is the geometrodynamic field. This is The Beauty in the Pattern.

Since $\Lambda > 0$ universally repels and $\Lambda < 0$ universally attracts we have the Bondi vacuum propeller in which the center of mass of the rigid device self-accelerates without any g-force, i.e. a self-generating time-like geodesic with on board Λ-field generators.

Flying Saucers, Black Ops, Hunt For The Zero Point Farce
——-Original Message——-
From: web62299@stardrive.org [mailto:web62299@stardrive.org]
Sent: Friday, September 06, 2002 11:53 AM
To: Jack Sarfatti;
Subject: Latest Haisch, Rueda, et al paper

http://xxx.lanl.gov/abs/gr-qc/0209016

"A possible connection between the electromagnetic quantum vacuum and inertia was first published by Haisch, Rueda and Puthoff (1994). If correct, this would imply that mass may be an electromagnetic phenomenon and thus in principle subject to modification, with possible technological implications for propulsion. A multiyear NASA-funded study at the Lockheed Martin Advanced Technology Center further developed this concept, resulting in an independent theoretical validation of the fundamental approach (Rueda and Haisch, 1998ab). Distortion of the quantum vacuum in accelerated reference frames results in a force that appears to account for inertia. We have now shown that the same effect occurs in a region of curved space-time, thus elucidating the origin of the principle of equivalence (Rueda, Haisch and Tung, 2001). A further connection with general relativity has been drawn by Nickisch and Mollere (2002): zero-point fluctuations give rise to space-time micro-curvature effects yielding a complementary perspective on the origin of inertia. Numerical simulations of this effect demonstrate the manner in which a massless fundamental particle, e.g. an electron, acquires inertial properties; this also shows the apparent origin of particle spin along lines originally proposed by Schroedinger. Finally, we suggest that the heavier leptons (muon and tau) may be explainable as spatial-harmonic resonances of the (fundamental) electron. They would carry the same overall charge, but with the charge now having spatially lobed structure, each lobe of which would respond to higher frequency components of the electromagnetic quantum vacuum, thereby increasing the inertia and thus manifesting a heavier mass."

It's not a good paper IMO. It is definitely badly written to say the least. I already read it.[111] What do you think I have been alluding to this past week?

Bill Unruh already trashed a similar paper of theirs on the "Rindler flux".

This paper makes many claims without detailed justification and their physical model is ill-posed, fragmentary and apparently self-contradictory though its hard to tell because their discussion is badly written and vague. Their theory appears to violate local covariance. They make an interesting mass spectrum claim but do not provide enough detail to judge it.

"We have now shown that the same effect occurs in a region of curved space-time, thus elucidating the origin of the principle of equivalence"

That sentence is misleading. The initial promise, the whole idea of Haisch and Puthoff was motivated by the Andre Sakharov 1967 conjecture that somehow gravity is a collective electromagnetic phenomenon. Indeed Joe Firmage[112] gave Bernie Haisch several million dollars to set up CIPA in Palo Alto as a kind of pseudo respectable front organization to seduce traditional physicists into UFO research. Bernie finagled a respectable Board including my old Cornell math professor Wolfgang Rindler and a Canadian cosmologist-astrophysicist named Paul Wesson. Bernie actually made a fairly good website on UFOs that may still be up there. Bernie also edited the fringe paranormal UFO New Age healing Journal of Scientific Exploration and is quite chummy with Jacques Vallee the French UFO researcher, turned Silicon Valley investment banker, who worked closely with Ira Einhorn in the late 70's in the early days of the ARPA pre-internet. This was when Ira had ATT in Philly in his back pocket to help Jacques at "The Institute for the Future".

Joe gave Bernie a lot more money than to us at ISSO, which obviously created bad feelings at ISSO. Biases must be admitted.[113] The feeling was that ISSO was really doing what Joe wanted whilst Bernie was playing a silly subterfuge no-win pseudo-academic game trying to make UFO breakthrough propellantless propulsion (BPP) respectable in the mainstream physics community. Mark Millis of the NASA BPP team who Joe also funded bent over backwards not to mention UFOs, but the simple fact was many of their key people, like Al Holt, were prominent in UFO research and we insiders at ISSO knew what

---

[111] Indeed Bernie sent me a copy of his paper. When I told him candidly what I thought of it, his feathers were understandably ruffled and he wrote back to me on 8/21/02 about this book (excerpts from the less nasty parts ☺) "You are living in an electronic fiction based on nothing but arrogance and bombast. No one takes you seriously, your influence is zero, you have published nothing in years (if ever), … and that not-even-a-joke book of yours (to paraphrase Pauli) will never be foisted on the public by any publisher."

[112] A mover-shaker in the Dot.Com Bubble founding US Web who had a close encounter with Beings of Light, like the Mormon "Moroni" I am told, and who also read one of Bernie Haisch's zero point inertia papers that got him all fired up like Don Quixote reading books on Chivalry. Joe is also a direct descendent of the Mormon Guru, Brigham Young. Joe has his eyes on The White House. He is still in his 30's.

[113] I arranged for Creon Levit from NASA Ames to meet with Joe in 1999 who was looking for a science director for ISSO. Creon took the job and I was de-facto senior theoretical physicist. Joe promised us that ISSO "came first". Creon would not have taken on the job otherwise. Joe consistently broke his word on this, and other matters, secretly funding several parallel groups including Mark Millis of NASA BPP, mostly crackpot, like an "Ormus Powder" Lab in Budapest, not only in the eyes of ISSO, but even in the eyes of Bernie Haisch and Hal Puthoff. Joe also did not clear his public remarks on physics with any of us leading to embarrassment as when Brian Greene ("Elegant Universe") berated me at State of the World Forum in 1999 for letting Joe make crackpot remarks on Einstein's relativity and spiral dynamics to the audience. Joe continues with the same silly rap even in 2002.

was really going on. The only relevant BPP person at CIPA was Giovanni Modanese for a short while.

But getting back to the physics, such as it is, in this paper: If one is to have a zero point vacuum energy electromagnetic theory of gravity with a view of getting a practical vacuum propeller[114] then one must show how Einstein's field equations emerge out of the SED[115] EM ZPE (zero point random noise). Of course they do not do that. I do. One must also show how to avoid the huge anti-gravity that the SED EM ZPE must make. This is the Λ problem. They do not do that either. I do. They do not even know this is a problem! Or, at least they never discuss it. Yet it is the key problem that invalidates their whole idea before they even get off home plate.

Hal Puthoff was a US Naval Officer before NSA and he had very high security clearances, as high as Kit Green's, (the psychiatrist bio-weapons expert in "Remote Viewers: The Secret History of America's Psychic Spies" by Jim Schnabel) at CIA and National Security Council at the time. Hal is also basically a straight arrow and probably has good reason to believe that flying saucers are really out there and that they have advanced technology that can kick our asses out of the sky if they wished to do so. The physics I am coming up with now seems to be confirming that. Indeed, that is why I have focused in on this problem since 1999 and especially now post 911 since the basically microwave technology may not be that difficult. Indeed the people at SARA we worked with at ISSO not only think that the 1943 Philadelphia Experiment was real in a significant sense but also, consistent with Nick Cook's generally technically silly "Hunt For The Zero Point", the SARA people have evidence that the Nazis really may have stumbled into something important in regard to propellantless propulsion. I cannot evaluate their beliefs really. What is important is that the former SARA Chief Scientist James Corum went to the Institute of Software Research in Fairmount, West Virginia to work on electromagnetic stress propulsion. ISR is a pet project of Senator Robert Byrd. The latest ISR webpage has been completely changed hiding the photo of Dr. Corum and his EM Stress Propulsion Project that was on the original. Dr. Corum is also an expert on Nicola Tesla and Gabriel Kron and has had access to the complete Tesla Archive in Beograd even under the former Serbian government that committed war crimes. Corum has published on the Philadelphia Experiment and thinks Jacques Vallee is way off base in his article on that alleged incident published by Bernie Haisch in the Journal of Scientific Exploration. Cal State Professor James Woodward and independently Hal Puthoff have shot down Corum's particular proposal, which may explain why it is not on the new ISR web page? Woodward is a strong opponent to Bernie Haisch's theory[116] even though he and Puthoff agree about Corum's proposal. Indeed the ISSO million dollar project with SARA on essentially the Corum idea proved negative. Creon Levit found an experimental error in the engineering we paid for that gave false positive results. That is not the only "quasi-Black" project we at ISSO had down there. The other one was apparently partly successful, but is dead in the water since Joe Firmage lost his money.

---

[114] That is Hal Puthoff's obsessive dream for more than 25 years since his tenure at the National Security Agency with microwave engineering genius Ken Shoulders who had to tell Hal to shut up about flying saucers at least till after lunch!

[115] Stochastic electrodynamics, a failed scheme to avoid quantizing the electromagnetic field directly which is part of Haisch's theory that does not, in my opinion, have the slightest chance of ever working as advertised.

[116] Bernie calls both me and Professor Woodward "nasty" for giving our honest opinions on his shoddy theory.

"Distortion of the quantum vacuum in accelerated reference frames results in a force that appears to account for inertia. We have now shown that the same effect occurs in a region of curved space-time, thus elucidating the origin of the principle of equivalence (Rueda, Haisch and Tung, 2001). A further connection with general relativity has been drawn by Nickisch and Mollere (2002): zero-point fluctuations give rise to space-time micro-curvature effects yielding a complementary perspective on the origin of inertia."

These claims are not justified IMO. Indeed, I think Bill Unruh wrote a refutation on Usenet on the "Rindler flux" claim as violating the equivalence + tensor covariance principles.

"we suggest that the heavier leptons (muon and tau) may be explainable as spatial-harmonic resonances of the (fundamental) electron"

They never explain this in any detail. What is resonating? — the quantum wave function? What? How does it self-trap? They also never explain what the massless electron actually is. They do not use wave functions of the electron in their math. They appeal to de Broglie and presumably his assistant of 30 years J. P. Vigier[117] which need a pilot wave and a particle but do not explain their physical picture here at all relying only on formalism without informal definitions. That goes against the de Broglie idea and since they have no electron wave functions, they really do not have a coherent theory in this paper. Haisch's equation (11) is

$$\Gamma^\mu_{\nu\rho} p^\nu p^\rho = -\frac{q}{c} F^\mu_\nu p^\nu n^0 \tag{5.1}$$

where Haisch's equation (10) is

$$n^\mu = \left(n^0, \vec{n}\right) \tag{5.2}$$

is not covariant as shown below in detail. Haisch's equation (15)

$$g = C \cdot \eta \cdot C \tag{5.3}$$

is not explained adequately. It is not a local tetrad map from LNIF to LIF at point event P that it formally resembles. It pretends to be a kind of Yilmaz transformation from a variable Curve World to a global Flat World. In contrast, the local tetrad map at P has the same local form but the map is from a point in the base space manifold of the tangent bundle to the local flat tangent space. In addition they never face up to the space-time stiffness problem or to the Λ problem.

---

[117] Bernie basically finagled Professor Vigier, over 80 years old, to sign his name to one of his papers that Vigier did not really have much, if anything, to do with. This upset the core people at ISSO especially Vladimir Poponin from Moscow who brought Gennady Shipov to ISSO from Moscow. This happened in the summer of 1999 when Vigier was living closely with ISSO Staff for two months in a Telegraph Hill Apartment I arranged for. Vigier was very uneasy at his first meeting with Joe Firmage. Joe did not listen but only spouted his crackpot views on Einstein's relativity. Vigier considered flying back to Paris the next day but Jagdish Mann calmed him down. All the core people at ISSO were aware of this.

Local Covariance as Form Invariance of The Laws of Physics

What does it mean? Let's take an example – Einstein's 1915 local geometrodynamic field equation of the gravitational field as space-time curved by source mass-energy density. Local means at a point P in space-time, i.e. where and when. Einstein's tensor equation is

$$G_{\mu\nu} = -\frac{G}{c^4} T_{\mu\nu} \qquad (6.1)$$

Define the local 1-1 reversible holonomic coordinate chart transformation symbols in a neighborhood of a fixed space-time point event P as[118]

$$x^{\mu'} = x^{\mu'}\left(x^{\mu}\right)_P$$
$$x^{\mu} = x^{\mu}\left(x^{\mu'}\right)_P \qquad (6.2)$$
$$\mu' \& \mu = 0,1,2,3$$

$$X_{\mu}^{\mu'}(P) \equiv \frac{\partial x^{\mu'}}{\partial x^{\mu}}\bigg|_P$$

$$(6.3)$$

$$X_{\mu'}^{\mu}(P) \equiv \frac{\partial x^{\mu}}{\partial x^{\mu'}}\bigg|_P$$

$$X_{\lambda}^{\mu'} X_{\nu'}^{\lambda} = \delta_{\nu'}^{\mu'}$$

$$(6.4)$$

$$X_{\lambda'}^{\mu} X_{\nu}^{\lambda'} = \delta_{\nu}^{\mu}$$

Note also the inhomogenous term in the parallel transport non-tensor connection field $\Gamma_{\mu\nu}^{\lambda}$ Diff(4) transform

$$X_{\nu'\lambda'}^{\mu}(P) \equiv \frac{\partial^2 x^{\mu}}{\partial x^{\nu'}\partial x^{\lambda'}}\bigg|_P \qquad (6.5)$$

These Diff(4) local transformations connect the observations of locally coincident observers Alice and Bob at P in arbitrary motion relative to each other. The term "locally coincident" is an approximate term meaning that the momentary space-time separation between Alice and Bob are small compared to the local radii of curvature at P. For example, the scale of the four radii of curvature of the Earth's source mass at a point P is of the order of 1 AU $\sim 10^{13}$ cm $\sim 10^3$ seconds. Therefore, Alice and Bob should not take their measurements more than 1000 seconds apart, but they can be quite far away from each other and still be coincident in the approximate sense meant here. The Earth's curvature at its surface is $\sim 10^{-26}$ cm^{-2}. This scale of curvature corresponds to $\sim 10^{92}$ Bekenstein-Hawking c-bits of black hole entropy. If a single electron is modeled as a sphere whose radius of

---

[118] Sum over repeated pairs of upper and lower tensor component indices. The Kronecker δ_{ν}^{μ} symbols are 1 for equal upper and lower indices (no summing) and 0 for different upper and lower indices.

curvature is h/mc ~ 10^{-11} cm. Its black hole entropy is ~ 10^{44} bits an Eddington number.[119] This requires an enormously strong short range Abdus Salam gravity constant G* in which

$$\frac{e^2}{mc^2} = \left(L_p^{*2} \lambda_c \right)^{\frac{1}{3}}$$

(6.6)

The M-theorists are playing with these much larger Planck lengths L_p^* from the unseen extra large curled up dimensions of hyperspace in modeling the many material brane worlds[120] of Super Cosmos all connected perhaps by Star Gate Time Travel Portals known as the "Underground Stream" to the Cabalist Alchemists. In modern jargon this is the Subway to the Stars and Beyond.

Einstein's field equation locally transforms as

$$G_{\mu\nu} = -\frac{G}{c^4} T_{\mu\nu} \rightarrow G_{\mu'\nu'} = -\frac{G}{c^4} T_{\mu'\nu'}$$

(6.7)

Where

$$T_{\mu'\nu'} \equiv X^{\mu}_{\mu'} X^{\nu}_{\nu'} T_{\mu\nu}$$
$$G_{\mu'\nu'} \equiv X^{\mu}_{\mu'} X^{\nu}_{\nu'} G_{\mu\nu}$$

(6.8)

Therefore

$$G_{\mu'\nu'} = -\frac{G}{c^4} T_{\mu'\nu'} \rightarrow X^{\mu}_{\mu'} X^{\nu}_{\nu'} G_{\mu\nu} = -\frac{G}{c^4} X^{\mu}_{\mu'} X^{\nu}_{\nu'} T_{\mu\nu}$$

(6.9)

Use (6.4) Kronecker orthonormality

$$G_{\mu\nu} = -\frac{G}{c^4} X^{\mu'}_{\mu} X^{\mu}_{\mu'} X^{\nu'}_{\nu} X^{\nu}_{\nu'} T_{\mu\nu} = -\frac{G}{c^4} \delta^{\mu}_{\mu} \delta^{\nu}_{\nu} T_{\mu\nu} = -\frac{G}{c^4} T_{\mu\nu}$$

(6.10)

Therefore the forms of the laws of nature are invariant under Einstein's local transformations between momentarily coincident observers Alice and Bob in arbitrary relative motion. They can have any order of relative accelerations of accelerations. Einstein's Equivalence Principle (EEP) is an additional natural condition that Bob is on a free float weightless time-like geodesic LIF whilst Alice is on a time-like non-geodesic LNIF[121] that approximately intersect narrowly missing a direct collision between Alice and

---

[119] This is an unexpected digression that emerged in my stream of consciousness ~ 7:40 PM PST, Sept 7, 2002 in San Francisco. A message from ET? Is this whole book a message from ET? ☺ You tell me.

[120] Piloted by the cosmic thoughts of Hawking's Mind of God.

[121] Alice feels weight from the electrical reaction forces that keep her off a time-like geodesic path. Haisch et-al claim that this effect is a drag force through the virtual photons, but only the transverse polarized ones. Their equations however are inconsistent not obeying local covariance.

Jack Sarfatti

Bob. This is the tetrad transformation $\xi_a^\mu(P)$ from the LNIF locally curved metric field $g_{\mu\nu}(P)$ to the locally flat tangent space metric η_{ab} where

$$g_{\mu\nu}(P) = \xi_\mu^a(P)\xi_\nu^b(P)\eta_{ab}$$
$$\eta_{ab} = \xi_a^\mu(P)\xi_b^\nu(P)g_{\mu\nu}(P)$$

(6.11)

Note whilst the flat metric η_{ab} has no local P dependence, the tetrad coefficients $\xi(P)$. This is precisely what Haisch & Co do not seem to understand in their eq. (15). They never indicate if their C is a local function or not? Indeed this so-called Nickisch-Mollere Connectivity theory is a patent Cargo Cult simulacrum, a Golem, of the well-known tetrad method that they do not mention.

Mathematical Inconsistencies in the Haisch Zero Point Inertia Theory

"For the record, in lectures I have given I always state that Rueda deserves most of the credit for the analysis. He is an expert in SED, got a Ph. D. from Cornell doing that, and has spent three decades doing SED-based research. He is a brilliant physicist; I'm really an astronomer (as you know). But I have initiated some of the basic concepts, such as the notion that the ZPF-matter interactions take place at a resonance rather than at some cutoff, the association of that resonance with the Compton frequency and the tie in of that to the origin of the de Broglie wave (inspired by Geoff Hunter by the way). I have often stated that I see myself as a catalyst for some potentially important ideas. I am satisfied with that role.

Are we right in our physics? Time will tell. To me it looks like a beautiful and elegant confluence of connections — especially given the new work by Nickisch and Mollere. But who knows? Not being the level of genius that you fancy yourself to be, I can't know yet.

Now, my friend, lets consider a couple of your claimed contributions to civilization's intellectual heritage. In your previous email you wrote:

Jack: The books I helped write did pretty well commercially. Space-Time and Beyond sold several hundred thousand copies. Dancing Wu Li Masters sold millions of copies and still is selling well - even though I was cheated out of my share of the royalties by Zukav.

Bernie: That's quite a claim since according to the book covers the author of Space-Time and Beyond was Bob Toben and the author of Dancing Wu-Li Masters was Gary Zukav. Are you claiming credit for their books? Are you accusing them of plagiarism? Was Jack Sarfatti the true author of both?

Better not to hurl accusations when you are in a China shop... especially one full of other people's china.

Bernie"

Dr. Bernard Haisch
519 Cringle Drive
Redwood Shores, CA 94065

———————————————————————

Chief Science Officer, ManyOne Networks, Inc.[122]
Director, California Institute for Physics and Astrophysics
Scientific Editor, The Astrophysical Journal
phone: 650-593-8581, fax: 650-595-4466
email: <haisch@calphysics.org>, alternate: <haisch@manyone.net>
http://www.calphysics.org

8/22/02 sent to an open public e-mail list.

Incompetent Error in "Update on an Electromagnetic Basis for Inertia, Gravitation, the Principle of Equivalence, Spin and Particle Mass Ratios" by Bernard Haisch, Alfonso Rueda, L. J. Nickish, Jules Mollere accepted for publication in AIP Conference on Space Technology (STAIF-2003) "Expanding the Frontiers of Space" Feb 2-6, 2003 Albuquerque

Peer Review by Jack Sarfatti, Ph.D. (Physics UC)

Incompetent Error "Update on an Electromagnetic Basis for Inertia, Gravitation, the Principle of Equivalence, Spin and Particle Mass Ratios" by Bernard Haisch, Alfonso Rueda, L. J. Nickish, Jules Mollere accepted for publication in AIP Conference on Space Technology (STAIF-2003) "Expanding the Frontiers of Space" Feb 2-6, 2003 Albuquerque

Their eq. (11), p. 5 is

$$\Gamma^{\mu}_{\nu\rho} p^{\nu} p^{\rho} = -\frac{q}{c} F^{\mu}_{\nu} p^{\nu} n^{0} \tag{7.1}$$

It is manifestly not locally covariant under Diff(4) general coordinate transformations. Therefore it violates Einstein's general theory of relativity and is not of the caliber of papers that should be published by the American Institute of Physics since the authors mistakenly think their calculations agree with Einstein's theory. Bill Unruh has raised a similar objection against another Haisch paper on "Rindler flux". For example, make a local Diff(4) transformation.

$$\Gamma^{\mu}_{\nu\rho} \to \Gamma^{\mu'}_{\nu'\rho'} = \Gamma^{\mu}_{\nu\rho} X^{\rho}_{\rho'} X^{\nu}_{\nu'} X^{\mu'}_{\mu} + X^{\lambda}_{\nu'\rho'} X^{\mu'}_{\lambda} \tag{7.2}$$

$$n^{0} \to n^{0'} \stackrel{?}{=} n^{0} X^{0'}_{0} \tag{7.3}$$

(7.3) violates local covariance. We have in general

$$n^{\mu} \to n^{\mu'} = n^{\mu} X^{\mu'}_{\mu} = n^{0} X^{\mu'}_{0} + n^{1} X^{\mu'}_{1} + n^{2} X^{\mu'}_{2} + n^{3} X^{\mu'}_{3} \tag{7.4}$$

Therefore

---

[122] ManyOne Networks is Joe Firmage's latest attempt to recoup his lost fortune apparently giving up on Motion Sciences, which never had a chance of flying in my opinion since it was entirely based upon the Haisch-Puthoff type models.

$$n^0 \to n^{0'} = n^\mu X_\mu^{0'} = n^0 X_0^{0'} + n^1 X_1^{0'} + n^2 X_2^{0'} + n^3 X_3^{0'} \tag{7.5}$$

The LHS of Haisch equation (11) (our 7.1) transforms as

$$\Gamma_{\nu\rho}^\mu p^\nu p^\rho \to \Gamma_{\nu'\rho'}^{\mu'} p^{\nu'} p^{\rho'} = \left(\Gamma_{\nu\rho}^\mu X_{\rho'}^\rho X_{\nu'}^\nu X_\mu^{\mu'} + X_{\nu'\rho'}^\lambda X_\lambda^{\mu'}\right) X_\nu^{\nu'} p^\nu X_\rho^{\rho'} p^\rho \tag{7.6}$$

The RHS of equation (11) transforms as

$$-\frac{q}{c} F_\nu^\mu p^\nu n^0 \overset{?}{\to} -\frac{q}{c} F_{\nu'}^{\mu'} p^{\nu'} n^{0'}$$

$$= -\frac{q}{c} F_\nu^\mu X_{\nu'}^\nu X_\mu^{\mu'} p^\lambda X_\lambda^{\nu'} \left(n^0 X_0^{0'} + n^1 X_1^{0'} + n^2 X_2^{0'} + n^3 X_3^{0'}\right)$$

$$= -\frac{q}{c} F_\nu^\mu X_{\nu'}^\nu X_\lambda^{\nu'} X_\mu^{\mu'} p^\lambda \left(n^0 X_0^{0'} + n^1 X_1^{0'} + n^2 X_2^{0'} + n^3 X_3^{0'}\right) \tag{7.7}$$

$$= -\frac{q}{c} F_\nu^\mu \delta_\lambda^\nu X_\mu^{\mu'} p^\lambda \left(n^0 X_0^{0'} + n^1 X_1^{0'} + n^2 X_2^{0'} + n^3 X_3^{0'}\right)$$

$$= -\frac{q}{c} F_\nu^\mu X_\mu^{\mu'} p^\nu \left(n^0 X_0^{0'} + n^1 X_1^{0'} + n^2 X_2^{0'} + n^3 X_3^{0'}\right)$$

If Haisch's (11) is to be covariant, then (7.6) and (7.7) must be equated. Note that I have used

$$X_{\beta'}^\alpha X_\gamma^{\beta'} = \delta_\gamma^\alpha$$
$$X_\beta^{\alpha'} X_{\gamma'}^\beta = \delta_{\gamma'}^{\alpha'} \tag{7.8}$$

on the RHS. Using it again on the LHS gives

$$\left(\Gamma_{\nu\rho}^\mu X_{\rho'}^\rho X_{\nu'}^\nu X_\mu^{\mu'} + X_{\nu'\rho'}^\lambda X_\lambda^{\mu'}\right) X_\sigma^{\nu'} p^\sigma X_\kappa^{\rho'} p^\kappa$$

$$= \left(\Gamma_{\nu\rho}^\mu X_{\rho'}^\rho X_{\nu'}^\nu X_\mu^{\mu'} X_\sigma^{\nu'} X_\kappa^{\rho'} + X_{\nu'\rho'}^\lambda X_\lambda^{\mu'} X_\sigma^{\nu'} X_\kappa^{\rho'}\right) p^\sigma p^\kappa$$

$$= \left(\Gamma_{\nu\rho}^\mu X_{\rho'}^\rho X_\kappa^{\rho'} X_{\nu'}^\nu X_\sigma^{\nu'} X_\mu^{\mu'} + X_{\nu'\rho'}^\lambda X_\lambda^{\mu'} X_\sigma^{\nu'} X_\kappa^{\rho'}\right) p^\sigma p^\kappa$$

$$= \left(\Gamma_{\nu\rho}^\mu \delta_\kappa^\rho \delta_\sigma^\nu X_\mu^{\mu'} + X_{\nu'\rho'}^\lambda X_\lambda^{\mu'} X_\sigma^{\nu'} X_\kappa^{\rho'}\right) p^\sigma p^\kappa \tag{7.9}$$

$$= \Gamma_{\nu\rho}^\mu \delta_\kappa^\rho \delta_\sigma^\nu X_\mu^{\mu'} p^\sigma p^\kappa + X_{\nu'\rho'}^\lambda X_\lambda^{\mu'} X_\sigma^{\nu'} X_\kappa^{\rho'} p^\sigma p^\kappa$$

$$= \Gamma_{\nu\rho}^\mu X_\mu^{\mu'} p^\nu p^\rho + X_{\nu'\rho'}^\lambda X_\lambda^{\mu'} X_\sigma^{\nu'} X_\kappa^{\rho'} p^\sigma p^\kappa$$

Therefore

$$\Gamma_{\nu\rho}^\mu X_\mu^{\mu'} p^\nu p^\rho + X_{\nu'\rho'}^\lambda X_\lambda^{\mu'} X_\sigma^{\nu'} X_\kappa^{\rho'} p^\sigma p^\kappa$$

$$= -\frac{q}{c} F_\nu^\mu X_\mu^{\mu'} p^\nu \left(n^0 X_0^{0'} + n^1 X_1^{0'} + n^2 X_2^{0'} + n^3 X_3^{0'}\right) \tag{7.10}$$

This consequence of the Haisch paper is obvious nonsense. That is, the form of their (11) is not invariant. In other words (7.1) above does not transform to

$$\Gamma^{\mu'}_{v'\rho'} p^{v'} p^{\rho'} = -\frac{q}{c} F^{\mu'}_{v'} p^{v'} n^{0'} \tag{7.11}$$

Conceptual Inconsistencies in Puthoff's Challenge to Einstein's Curved Space-Time Explanation of Gravity

Puthoff's key PV paper upon which I base my refutations is at http://xxx.lanl.gov/abs/gr-qc/9909037 . Puthoff claims to have secret flying saucer work that Eric Davis alludes to in http://198.63.56.18/pdf/davis_mufon2001.pdf on the applications to UFOs that I have not seen. However, I do not believe his claims since his basic idea is not even wrong in my opinion.

In a message dated 9/06/02 5:41:50 PM to an *open list*, sarfatti@well.com writes:

<< Dicke did it 42 years ago. You simply copied Dicke's work on this. >>

Though not yet published, I've gone way, way beyond Dicke. Application to cosmology, application to charge cluster phenomena[123], application to energy/propulsion, application to ....

Collegially,
Hal

In a message dated 9/04/02 3:29:48 PM, sarfatti@well.com writes:

<< I am also sending this to Hal Puthoff in case he is smart enough to jump on my Band Wagon >>

Sorry, but the tune is discordant with my melodious Lagrangian! :-) I realize you can't hear my tune from inside the middle of your band, so no blame, it's a no-fault condition!

Hal

In a message dated 8/30/02 10:13:38 AM, sarfatti@pacbell.net writes:

<< Now Hal Puthoff claims that his real K PV theory is secret and what he publishes is not really it. >>

---

[123] Experimental work of Ken Shoulders that Ron Pandolfi suspects is a spurious effect of charged mercury liquid drops. This is out of my field of expertise.

What I actually claim is that what I've published is only part of it, not that it's not really it. There is a difference. It's just that there is a lot of unpublished material and its implications, not yet published.[124]

Hal Puthoff
In a message dated 8/28/02 11:07:22 PM, sarfatti@pacbell.net writes:

<< However, I added that summing an infinite number of diagrams is a qualitative phase transition and you can no longer use a flat background picture....... In any case it has nothing to do with your PV which is completely classical. You are not summing an infinite number of spin 2 graviton Feynman diagrams >>

Puthoff: You missed the point.

The point is that one can begin with a flat background and carry out a
quantum sum to result in nonlinear GR equations with the flat background unobservable (as you describe above);

Or one can begin with variable vacuum refractive index parameters and carry out a classical derivation to result in nonlinear GR equations with the flat background unobservable.

As in Kittel and other standard texts, one can treat material properties such as permittivity either classically or quantum mechanically, depending on the purpose of the derivation.

To paraphrase Paul Zielinski's quote, "Einstein's cat can be skinned many ways." MTW gives six in Box 17.2 ("Six routes to Einstein's Geometrodynamic Law ....." PV is a variation on this theme.

<<Jack: Not only that, I get the new vacuum propeller $g_{\mu\nu}\,\partial\Lambda/\partial x^{\nu}$ zero point term[125] without which you cannot do anything >>

The corresponding Λ - type term in PV is of course a function of the vacuum dielectric constant K.

Collegially,
Hal Puthoff

In a message dated 8/29/02 5:11:14 PM, sarfatti@pacbell.net writes:

<<Jack: Do you agree for the record that this is my original idea?>>

Hal: Yes, the double boundary layer of dark energy/dark matter is definitely your idea. I'm glad to tell anyone that.
    Jack: Do you have anything better? I will eat my hat if you do.
    Hal: Better as a theoretical construct? Mmmm, probably not. Better as an implementable, engineering strategy? Possibly, it's too early to tell. Since the mechanics I

---

[124] Promises, promises. Talk is cheap, it's the equations that count here. God is in the details.
[125] In the generalized Bianchi identity for local momenergy current density conservation.

am looking at derive from the PV approach, you don't have to worry about it as competition since it is clearly "not even wrong!"

Jack: Do you agree that Iranian, Iraqi and mainland Chinese physicists are quite competent in this arena and are able to develop this with relatively small budgets?

Hal: Absolutely.

Jack: OK that is something we agree on 100%. That's important.

Hal: We're (hopefully!) on the same team. Think back to our collaborative discussion of Wesson's "gravitational fine structure constant" and Gray's Navy material, etc.

Jack: ......In any case you admitted you had no clue how to cohere the zero point — that's the key and that's what I have solved.

Hal: To my way of thinking, changing vacuum parameters (i.e., modulating K) is a form of modifying the underlying ZPE (which could be seen as a form of

In a message dated 8/27/02 6:22:32 PM, sarfatti@well.com writes:

<< Jack: No Tony you are wrong about your last statement. You are way behind the times. See the book "The Geometric Universe" for example. Perturbation theory is only in weak field limit. >>

Puthoff: No, Jack. Read MTW. The "perturbative" approach sums to the full nonlinear Einstein equations.[126]

Hal

Thanks for the references, Tony. And of course you make a good point!

And, as with one of the quotes, the "flat-space background" is non-observable[127] in our PV theory (because of rod and clock distortions) as is the case in the example you quote. Same, same.

Best regards,
Hal Puthoff

In a message dated 9/06/02 1:22:41 AM, sarfatti@well.com writes:

<< What does Hal actually have?>>

Hal: A terrific engineering approach (metric engineering) for applications of GR. :-)

<<He has a K.>>

Hal: Placeholder for metric tensor components

---

[126] This is a Red Herring since I never denied that. Indeed, just the opposite. I made that point initially.

[127] Therefore "excess metaphysical baggage" violating Einstein's Equivalence Principle in which the LIF local flat tangent space-time is observable. The global Yilmaz flat space time is simply not there. It was destroyed in the vacuum phase transition from Flat World to Curve World in forming each "Champagne bubble" brane world in Super Cosmos of hyperspace with hidden curled up extra dimensions. Note that Rabbi Solomon ha-Zarfati was a vintner in Champagne, France (1040 – 1105) at the time of The First Crusade to find The Temple in Jerusalem.

Jack: Nonsense. Only for SSS[128] case and even then it gives stupid answers. Even if so, so what? What's your point? Who do you think will be convinced? Convinced about what?

<<He has a classical differential equation for it....>>

Hal: As did Einstein.

Jack: So what? The point is that I have a macro-quantum correction term that explains new observations. Also I derive Einstein's equation from the macro-quantum theory. Hal, surely you are joking here? You do not seem to take your own ideas seriously enough to mount an even plausible to the point defense.

<<...that comes from a dubious action.....>>

Hal: that leads to the results of the standard classical tests of GR. BTW, where is your action principle? - - I haven't seen any Lagrangian in your work lately.

Jack: Formulating the action is not necessary for the validity of the dynamics. Newton did not have an action for his 3 laws. All that came much later. One can point to the Modanese action as a first step. My theory is more complicated as it has the feedback loop in which Ψ creates $g_{\mu\nu}$ & Λ, which in turn shapes Ψ. Some smart mathematician will figure out the action. Where is my David Hilbert? Einstein did not figure out his own action either. That is only a formalistic nicety. My theory is so nonlinear it is hard to write the action off the top of one's head. My current inability do so does not impact on the physical explanation and predictions I am getting any more than Newton's not having an action in sense of modern calculus of variations did not come until what 100 years after his death? You have an action, which you stole from Dicke, and what good has it done you in 15 years? Not much. When I have time I will probably figure out the action. I have not worked on it seriously yet. I am too busy with the physical ideas that are very superficial in your own approach. Getting agreement with 3 classic tests of GR is no big achievement on your part BTW. I trivially get that since I derive

$$G_{\mu\nu} + \Lambda g_{\mu\nu} = -\frac{8\pi G}{c^4} T_{\mu\nu} \text{ from } \Psi \text{ using Kleinert's method!}$$

<<.... with the variable c. >>

Hal: as in Einsteinian theory, c' = c[1 + 2 phi/c^2] for weak fields.

Jack: Big deal. Besides you did not really do it at all. Dicke did it 42 years ago. You simply copied Dicke's work on this.

Collegially,

Hal

From: Paul Zielinski [mailto:pzielins@ix.netcom.com]

---

[128] Spherically Symmetric Static Space-time in which a global set of LNIF rest frames relative to the source mass bending space-time into curve world can be used..

Sent: Thursday, September 05, 2002 10:21 PM
To: sarfatti@well.com
Subject: Re: Hal Puthoff & USAF Next Force

Jack,

Yes, you're right. I didn't understand that you were recovering the GR metric field out of a 2-fluid BE condensate.

I am looking at your papers carefully now to see if I can get a grip on this.

Paul

"if we remove life from Einstein's beautiful theory by steam-rolling it first to flatness and linearity, then we shall learn nothing from attempting to wave the magic wand of quantum theory over the resulting corpse." Roger Penrose, FRS, Professor of Mathematics at Oxford University

Bernie Haisch, Hal Puthoff and Joe Firmage have not yet learned that the world is not fundamentally "flat, stale and unprofitable."

For more details see Abhay Ashtekar's "Geometric Issues in Quantum Gravity" in "The Geometric Universe" Oxford, 1998.

What does Hal actually have?

PV without PV is to Einstein's curved space-time what Hollow Earth theory is to space science.

He has a K.

He has a classical differential equation for it that comes from a dubious action with the variable c.

In the SSS he shows, actually Dicke showed 42 years ago, that

$K = \exp[2GM/c^2r]$

Formally solves the differential equation.

However, a real PV theory must have some quantum physics in it or else it is false advertising.

I see no h in Hal's papers on K.

In contrast I have a macro-quantum coherence theory for both Einstein's geometrodynamic field $g_{\mu\nu}$ and the new quintessent field Λ now actually observed.

Hal is not even close. Look closer at his K. It's like the end of The Wizard of Oz, move the veil aside, what do you see?

Smoke and Mirrors in the Hunt for the Zero Point Farce.

Some useful links

On Dirac's ideas:

http://www.pparc.ac.uk/frontiers/current/feature4.asp

http://www.stardrive.org/Jack/dirac.pdf

(primitive ideas of mine in 1999-2000 at ISSO with J. P. Vigier with us for 2 months in San Francisco's North Beach).

http://arxiv.org/PS_cache/quant-ph/pdf/9608/9608024.pdf

This paper by David Finkelstein also independently has "vacuum as condensates". I just found it now for first time. Of course I brought Fink to Esalen in 1976. I knew him from Yeshiva University in the mid 60's with Lenny Susskind. I edited in Fink's "quantum logic of non-Boolean lattices of yes-no quantum propositions for Gary Zukav in Dancing Wu Li

Masters. Gary could not pass a course in elementary algebra. About binomial theorem Gary was not exactly teeming with a lot of news and he knew nothing much about the square of the hypotenuse (Einstein's relativity).[129] ;-)

Fermi sphere micro-quantum vacuum:

http://www.physik.unistuttgart.de/ExPhys/2.Phys.Inst./official/m.mehring/CondMat/densityofst.htm

http://www-user.tu-chemnitz.de/~sol/md/stab/node4.html

http://www.physics.wustl.edu/~visser/Analog/CollectiveModes.eps

Energy density of the micro-quantum vacuum, i.e. the Λ problem of "The Unbearable Lightness of Being":

Read this by John Baez and see how lame Hal's explanation is for this. He doesn't even have a fig leaf to hide behind!

http://math.ucr.edu/home/baez/vacuum.html

http://www.lns.cornell.edu/spr/2002-02/msg0039483.html

Quantum field theory

http://www.mit.edu/people/kerson/c0ntents.htm

Nonlocality vs macro-quantum superfluids

http://www.fdavidpeat.com/bibliography/essays/healtech.htm

Bose-Einstein condensates

http://amo.mit.edu/~bec/intro/whatbec/whtisbec.html

PPS Paul - also you are way off the mark. I have a really detailed theory of the physical vacuum Hal has nothing but a classical black box K with no theory explaining K! I have the details inside the box! You obviously did not understand my idea of the unstable random noisy high entropy choppy pre-gravity micro-quantum Dirac Fermi Sphere Flat World Vacuum of Hal and Bernie collapsing into the partially coherent (two-fluid) stable smooth quieted non-random zero initial entropy macro-quantum BCS paired virtual electron-positron pairs in bound state Bose-Einstein condensate whose coherent phase modulation IS Einstein's Curve World $g_{\mu\nu}$ and whose amplitude is ~ Λ the quintessent zero point random residual field left over After The Fall.

$\Lambda > 0$ = dark energy macro-quantum vacuum region with universal repulsion

(residual random virtual boson dominated)

$\Lambda < 0$ = dark matter macro-quantum vacuum region with universal attraction (residual random virtual unbound ionized plasma electron-positron pairs)

Ergo, Bondi Vacuum Propeller Weightless Warp Drive on small scale + Star Gate Time Travel.

What does Hal Puthoff have to compare as a fundamental theory and heuristic for physical vacuum? Simple he's got nothing! From nothing comes nothing!

Therefore Paul your remark:

Z: *"but I also see the value of physical vacuum approaches such as Hal's version of Dicke's PV in the longer term, since they have the *potential* to lead to heuristically more powerful theories at a deeper level of physical understanding"*

Is completely unjustified by the facts of Hal's dilute disquisitions.

Furthermore your remarks:

Z: *"and may eventually trigger a fundamental shift out of the GR metric field paradigm if coupled with a parallel return to concrete models in QM. ... . My own view, for what it's*

---

[129] Major General's Song, Pirates of Penzance, Gilbert and Sullivan.

worth, is that GR (notwithstanding all the pseudo-positivistic curved-space PR) is really a physical field theory dressed up in metric clothing, with real physical gravitational fields and forces — which IMO lends considerable plausibility to Yilmaz's basic idea."

Jack: Yilmaz is to Einstein, what David Irving is to Stephen Schwartz! — or what Nick Herbert is to me. This reminds me of Nick Herbert talking about Israel and Palestine citing Max Weiss and David Irving.

http://www.ladyofthecake.com/mel/prod/sounds/springti.wav
http://www.tmbhs.com/tmbhs/movies/theproducers/theproducers.asp
http://qedcorp.com/book/psi/hitweapon.html

The Yilmaz theory is not the SSS theory; it is the SS Einstein-bashing theory from Himmler's Boys in the 1930's. So is that Ormus Powder with supposedly anti-gravity powers that Joe put so much moolah into in Transylvania! ☺

Finally your remark

Z: "Whether his proposed modification of the GR field equations is actually viable is another question."

Jack: I show below in detail why PV is not viable. It's simply more smoke and mirrors on the level of Nick Cook's tome of disinformation "The Hunt for the Zero Point" Farce that will rank with "Majestic 12 UFO Papers", "Elders of the Protocols of Zion", and Gary Zukav's shabby New Age self-help Me-Generation psycho-babble est-speak "De Seat of Da Sold" as classic cases of dis and mis information. ☺

PZ wrote: They say that this "effective" field stress-energy is contained within a "macroscopic region" of the order of several gravitational wavelengths.

I answered: OK, but no one disputes that say gravity waves propagate energy. The dispute is over the existence of a local stress-energy density tensor $t_{\mu\nu Yilmaz}$ in vacuum, which would replace

$$R_{\mu\nu}(P) = 0 \qquad (7.12)$$

by a purely classical non-quantum

$$G_{\mu\nu}(P)_{vac} \equiv R_{\mu\nu}(P) - \frac{1}{2} R_{\lambda}^{\lambda}(P) g_{\mu\nu}(P) = -\frac{8\pi G}{c^4} t_{\mu\nu}(P)_{Yilmaz} \qquad (7.13)$$

Note, my nonclassical net locally random micro-quantum zero point fluctuation correction term is of this generic form with additional (non-Yilmaz) condition that

$$\frac{-8\pi G}{c^4} t_{\mu\nu}(P) \equiv \Lambda(P) g_{\mu\nu}(P) \qquad (7.14)$$

where

$$\Lambda(P) \equiv L_p^{-2} \left(1 - L_p^3 \left| \Psi(P)_{MACRO} \right|^2 \right) \qquad (7.15)$$

$\Psi(P)_{MACRO}$ = locally variable virtual bound state e+e- macro-quantum Bose-Einstein condensate where[130]

$$g_{\mu v}(P) = \eta_{\mu v} + \frac{L_p^2}{2}\left(\frac{\partial}{\partial x^\mu}\frac{\partial}{\partial x^v} + \frac{\partial}{\partial x^v}\frac{\partial}{\partial x^\mu}\right)\arg\Psi(P)_{MACRO} \qquad (7.16)$$

= Elastic strain tensor of Hagen Kleinert's "World Crystal" Planck Lattice where $\arg\Psi_{MACRO}$ string singularities of disclination are curvature and of dislocation are torsion.

$$G_{\mu v} \to R_{\mu v} = 0, \Lambda \to 0 \qquad (7.17)$$

The generalized Bianchi identity corollary is the vacuum propeller equation for time-like geodesic warp drive (Paul Hill's "acceleration field" for the flying saucer in "Unconventional Flying Objects")

$$G_{\mu v}^{;v} + \frac{\partial\Lambda}{\partial x_v}g_\mu^v \approx 0 \qquad (7.18)$$

Where Λ has a U(1) gauge invariant Josephson effect dependence on the EM 4-potential A_μ for a "weak link" between a real superconductor and the vacuum superconductor.

$\Lambda > 0$ is universal anti-gravity "dark energy" in space-time neighborhood of P. This is a gravity blue shift source, e.g. gamma burster?

$\Lambda < 0$ is universal gravity "dark matter" in space-time neighborhood of P. This is a gravity red shift source - Arp's close quasars with anomalous redshifts?[131]

Einstein vacuum is non-gravitating $\Lambda = 0$.

The flat plate Bondi capacitor vacuum propeller self-propels along a time-like geodesic along the $\Lambda > 0 \to \Lambda < 0$ direction (to the right in this example). There is no limit to the effective speed reached as seen by outside observers because internally the crew is "not moving at all"! The effective warp speed can be many times c. There are no g-forces and the tidal forces can be kept to a minimum.

On 9/5/02 9:22 AM, "Puthoff@aol.com" <Puthoff@aol.com> wrote:

"I told you that before, but you missed it. You want and need to see what I'm doing as "not even wrong." Too bad, our insights might very well complement each other if dialog could replace polemics. But your dismissive approach towards my approach does not inspire me to share beyond what I've already done. You've insisted that I at least share with Paul or Creon. I have shared with Paul, he does get it, and you trash his attempt to get you to see the infrastructure forest, not just a few trees that have your name on them. Too bad."

Jack: I think you are fooling yourself. I have seen nothing from Paul, which fits what you just said.

---

[130] This formula corrects a systematic typo error in my previous book *Destiny Matrix* that left out the flat metric term, which could not be corrected in time because the file was so large 360 megs that any change might have corrupted it.

[131] The pulsating red and blue shifts allegedly seen from flying saucers reported by Eric Davis below suggest a Λ field oscillation.

Paul Z: "For the record, I see the force of Jack's general arguments on behalf of his theory in terms of shorter term progress, but I also see the value of physical vacuum approaches such as Hal's version of Dicke's PV in the longer term, since they have the *potential* to lead to heuristically more powerful theories at a deeper level of physical understanding, and may eventually trigger a fundamental shift out of the GR metric field paradigm if coupled with a parallel return to concrete models in QM."

I wrote: i.e. http://stardrive.org/Jack/PT81502.pdf

http://stardrive.org/Jack/nextforce.pdf

Are my latest expositions of my fast developing surprisingly rich fertile creative original theory with I think vital immediate implications for US Space Weapons Strategic Planning. If you think that, then explain what you think is right with Hal's PV that is 90% lifted from an old obsolete ~ 1960 Dicke paper in the period of GR Feynman, per his pattern of outrageous theatrics, called stupid whilst hissing like a rattle snake jumping up and down on the top of a table in the Cal Tech Cafeteria with hundreds of students spellbound. No one has addressed my several particular objections. Some of them are:

* He needs an infinite number of isotropic radial coordinates for each unique curvature coordinate! What's the Penrose diagram for that?

* All of his isotropic coordinates are complex when curvature coordinate $r_c < GM/c^2$ in his PV SSS model.

* In comparison, Einstein's GR has only 2 isotropic coordinates and they both have physical meaning in the Penrose diagram. Also they go complex when $r_c < 2GM/c^2$

* Hal's Tables I & II are inconsistent with his own math! In fact the rulers do shrink anisotropically in PV when you do it correctly - just like in Einstein's GR (different radial dependence of course - the shrinkage is not isotropic like in the Tables). Hal did not do his own math correctly here!

* This is why I say in my book http://stardrive.org/Jack/cover.jpg that Hal does not know strong field from weak field. Hal has been out so long it looks like in to him.

* Hal's action is inconsistent with Flat World SR, which is a global theory.

* Hal's equations are in violation of EEP + local covariance. He can force the latter on them, but it is artificial and he loses the whole point of PV-Yilmaz in that case.

* Hal cannot solve the rotating disk problem. He hand waves it's easy - for years he hand waves.

* Hal cannot solve the Λ problem - he hand waves and his lame explanation is so bad I am embarrassed for him. Definitely not up to Cornell Physics standards ~ late 50's.

* Hal admits he is clueless about how to cohere the ZPE - that's the baby in the bathwater!

Paul Z: "Also, Hal's positioning of PV as a heuristic model for engineering purposes is orthogonal to Jack's positioning of his theory as fundamental theoretical physics.

Jack: Hogwash Paul. You have not understood http://stardrive.org/Jack/nextforce.pdf - redownload it maybe you have older version. My theory is superior to Hal's here on both fronts. Fundamental theory and emerging gedankenexperiments to develop into real experimental tests as well as explaining:

* Dark energy

* Dark matter

* Unbearable lightness of being, i.e. Λ paradox that is like the black body UV infinity in 1900 that led Max Planck reluctantly to his quantum of action h.

* Arrow of Time

    * Bondi Vacuum Propeller

    * Star Gate Time Travel To The Past, i.e. Back *from* the Future. That's the hard part.

Paul Z: "At the same time, however, Jack is claiming a number of specific lab-testable novel predictions, which IMHNPO is for now the bottom-line nub of the whole matter: heuristic power and heuristic potential."

Jack: You got that right Bhubba. We have no time to spare. I am like Evariste Galois[132] the night before The Duel. No time for Laputan Pedants[133] peering at the twigs on the ground when the "Tyger Tyger eyes burning bright in the darkness of the night"[134] is about to leap down from the highest branch of Dan Smith's Eschaton[135] and gobble them up for a Midnight Nosh!

Z: "As far as I can see the jury is still out on Yilmaz. My own view, for what it's worth, is that GR (notwithstanding all the pseudo-positivistic curved-space PR) is really a physical field theory dressed up in metric clothing, with real physical gravitational fields and forces — which IMO lends considerable plausibility to Yilmaz's basic idea. Whether his proposed modification of the GR field equations is actually viable is another question."

Jack: Paul you do not understand the ideas of differential geometry very well. Also you do not understand EEP as explained for example by Stephen Hawking on page 17, Fig 1.11 of "The Universe in a Nutshell". Sure you can say shrinking rods - but that IS curved space time expressed informally (Bohm) as Kip Thorne says - no significant difference.

The basic error Hal and Bernie make, seized by Joe Firmage like Joan of Arc seizing the Fleur-de-lis on her way to The Burning Woman Festival, is trying to make a tempest out of a teapot. The mad see distinctions that are not there and fail to see distinctions that really are there.

This is partly Niels Bohr's fault, i.e. "create your own reality" from a too simplified rendition of his Smoky Dragon[136] from his Foggy Philosofawzy.[137]

I want results. I want them fast. And, surprisingly, I am getting them!

Mathematical Inconsistencies in Hal Puthoff's Polarizable Vacuum Approach to General Relativity.

    The Hal Puthoff paper I refer to is online at

---

[132] A Mozartian Prodigy, the creator of finite group theory limits of solvability that anticipates Godel's incompleteness theorem and the Turing halting theorem. Killed at 19 in 1832 by Napoleon II's Secret Police for his communist politics.

[133] Gulliver's Travels by Jonathan Swift parodied in Princess Ida by Gilbert and Sullivan..

[134] William Blake allusion.

[135] Dan Smith, an eccentric Blue Blood millionaire member of the legendary UFO Aviary of Black Ops Urban Legend that includes Hal Puthoff. See Erik Davis's book Techgnosis for background. Dan's family included the Pasadena Throops who created Cal Tech. Dan is like the character in "The Ruling Class" who, like Joe Firmage, cuts a Messianic figure for The Second Coming. Unlike Joe, Dan has physics degrees from Princeton and Stanford. Dan blames his madness on me because he read Space-Time and Beyond as an impressionable young man. Joe has not offered any similar excuse. ☺

[136] Image used by John Archibald Wheeler to describe the Copenhagen Fairy Tale of wave without particle, or BIT WITHOUT IT saved by the Deux Ex Machina miracle of the Marxist-inspired Von Neumann "collapse" of the quantum state.

[137] Feynman's derogatory term for the Cornell Philosophy Department when he was there. A Laputan Pedant does Philosofawzy and is in John Nash's favorite put-down a "hack". ☺

http://arxiv.org/abs/gr-qc/9909037 "Polarizable Vacuum Approach to General Relativity". Oops that was a typo, it should be "Polarizable"! ☺

Dr. Eric Davis, formerly at Robert Bigelow's Las Vegas NIDS UFO R&D Think Tank[138] promotes Puthoff's theory as UFO-Star Gate Physics in

http://198.63.56.18/pdf/davis_mufon2001.pdf

See also "The Star Gate Conspiracy" by Picknett & Prince.

Hal's SSS vacuum metric is

$$ds^2 = \frac{1}{K}c^2 dt^2 - K\left[dr_i^2 + r_i^2 \left(d\theta^2 + \sin^2\theta d\phi^2 \right) \right]$$
$$K \equiv e^{2GM/c^2 r_i}$$

(8.1)

Hal is using the isotropic radial coordinate r_i with the usual spherical polar angle coordinates polar latitude and azimuthal longitude $\theta \& \phi$ respectively. This is for a special LNIF observer Alice on a time-like non-geodesic at point P that is at rest relative to the source mass M bending space-time. This can be done in a stationary metric but not so easily in non-stationary metrics from rotating masses and when gravity waves wiggling space-time are passing through. One must be more careful in those cases. Hal does not appear to really understand Einstein's Equivalence Principle (EEP) when he writes down

$$ds^2 = c^2 dt_0^2 - \left(dx_0^2 + dy_0^2 + dz_0^2 \right)$$

(8.2)

because when you look at his not even wrong Tables I & II he thinks of this flat metric nonlocally as only at space-like asymptotic infinity. Hal does not appear to understand that this metric applies the LIF observer Bob whose time-like geodesic path intersects Alice's LNIF time-like non-geodesic path at the point P with a given r_i. This terrible ambiguity is a fatal infection of Hal's model. One never sees the general coordinate transformations $X_\mu^{\mu'}(P)$ explicit in any of Hal's explanations. That is, his theory is neither locally covariant nor does it obey EEP. Hal never explains how it can be that Bob in free float through the dielectric vacuum has undistorted rods and clocks whilst Alice's rods and clocks at the same time and place distort!

Einstein's SSS vacuum metric[139] in the same local isotropic coordinates for LNIF Alice at rest relative to the source mass M is

---

[138] The NIDS Science Advisory Board included Hal Puthoff, Jacques Vallee, and Army Colonel retired John Alexander, a noted paranormal expert and at one time head of the nonlethal weapons at Los Alamos. Colonel Alexander wrote a famous paper in the 1980 Military Review on military uses of the paranormal. His most recent book is Future War. Bigelow, in effect the successor to Howard Hughes, has taken most of the money out of NIDS to use for Bigelow Aerospace to build space stations in the private sector. Puthoff did the CIA Remote Viewing work at Stanford Research Institute on Uri Geller, Pat Price and Ingo Swann in the early 70's with Russell Targ where I first met the both of them. See Martin Gardner's "Magic and Paraphysics" in "Science, Good, Bad and Bogus".

$$ds^2 = \left(\frac{1-\frac{GM}{2c^2 r_i}}{1+\frac{GM}{2c^2 r_i}}\right)^2 c^2 dt^2 - \left(1+\frac{GM}{2c^2 r_i}\right)^4 \left[dr_i^2 + r_i^2\left(d\theta^2 + \sin^2\theta d\phi^2\right)\right] \qquad (8.3)$$

This same metric field in the more usual curvature radial coordinate r_c is

$$ds^2 = \left(1-\frac{2GM}{c^2 r_c}\right)c^2 dt^2 - \frac{dr_c^2}{\left(1-\frac{2GM}{c^2 r_c}\right)} - r_c^2\left(d\theta^2 + \sin^2\theta d\phi^2\right) \qquad (8.4)$$

A metric theory by definition is a curved space-time theory, which is built up on the idea of the geodesic as the straightest path in a curved space-time. The curvature coordinate in any SSS vacuum theory is defined such that the area of a concentric sphere is the flat Euclidean area $4\pi r_c^2$. Hal inconsistently claims he has a metric theory that obeys EEP yet he says he does not need the idea of curved space-time. This is nonsense. Thinking of Hal's formal math as a metric theory nevertheless, the relation between the two kinds of radial coordinates, as shown by Hal's assistant, Michael Ibison, is for PV

$$r_c = \sqrt{K} r_i = e^{GM/c^2 r_i} r_i \qquad (8.5)$$

This is in contrast to Einstein's simpler theory in which

$$r_c = r_i\left(1+\frac{GM}{2c^2 r_i}\right)^2 \qquad (8.6)$$

This is a simple quadratic equation giving two isotropic roots $r_{i\pm}$ for a fixed curvature r_c. These two isotropic roots have important topological meaning for the complete curved vacuum manifold forming the singular non-traversable wormhole Einstein-Rosen bridge.[140] Hal seems blissfully ignorant of this kind of mathematics in his self-described "metric engineering approach" – so like an engineer![141] ☺ Hal's theory over-hyped by Nick Cook in "The Hunt for the Zero Point" is simpler than is possible. Brian O Leary's similar book is even worse as a source of disinformation and misinformation on zero point energy technology, which does not yet exist. It is easy to see that Hal's theory has an infinite number of isotropic roots for the same curvature radial coordinate, yet Hal irrationally insists that the isotropic coordinates are superior even though they go complex in the interesting "strong field" region. Hal is seemingly unaware that he has an infinite number of

---

[139] The Riemann curvature tensor reduces to the conformal invariant Weyl tensor in Einstein vacuum with $\Lambda = 0$. We have Wheeler's mass without mass in which the source mass M is really a topological wormhole or Einstein-Rosen Bridge with event horizons at which time stops for the outside observers.
[140] Indeed the two isotropic roots in Einstein's theory are coordinate patches that only partially cover the whole curved manifold outside the two wormhole mouth event horizons in parallel brane worlds of the Einstein-Rosen bridge. This is why Hal thinks mistakenly that he has no black holes in his theory.
[141] Says the snooty theoretical physicist! ☺

his precious coordinates that are complex, hence unphysical, in the "strong field". What Hal thinks of as "strong field" is actually "weak field". Hal has been out on his PV limb so long that it looks like in to him. The limb is breaking.[142] ☺

In more formal detail:

First note the limits in Einstein's theory

$$r_i \rightarrow \infty \Rightarrow r_c \rightarrow \infty$$
$$r_i \rightarrow 0 \Rightarrow r_c \rightarrow \infty \tag{8.7}$$

Where a vanishing isotropic radial coordinate is not the strong field Hal thinks it is, but is simply space-like asymptotic infinity in the parallel brane universe next door of the Einstein-Rosen Bridge. It is not a bridge really because you cannot cross it. The space-time singularity will stretch and squeeze you to death as long as you are on any time-like or light-like world line.

Next solve the quadratic equation for GR.

$$r_c = r_i \left(1 + \frac{GM}{c^2 r_i} + \left(\frac{GM}{2c^2 r_i} \right)^2 \right) = r_i + \frac{GM}{c^2} + \left(\frac{GM}{2c^2} \right)^2 \frac{1}{r_i}$$

$$r_c r_i = r_i^2 + \frac{GM}{c^2} r_i + \left(\frac{GM}{2c^2} \right)^2 \tag{8.8}$$

$$r_i^2 + \left(\frac{GM}{c^2} - r_c \right) r_i + \left(\frac{GM}{2c^2} \right)^2 = 0$$

$$r_i^2 + \left(\frac{GM}{c^2} - r_c \right) r_i + \left(\frac{GM}{2c^2} \right)^2 = 0$$

$$r_{i\pm} = \frac{\left(r_c - \frac{GM}{c^2} \right) \pm \sqrt{\left(\frac{GM}{c^2} - r_c \right)^2 - \left(\frac{GM}{c^2} \right)^2}}{2} \tag{8.9}$$

$$\sqrt{\left(\frac{GM}{c^2} - r_c \right)^2 - \left(\frac{GM}{c^2} \right)^2} = \sqrt{\left(\frac{GM}{c^2} \right)^2 - \left(\frac{GM}{c^2} \right)^2 + r_c^2 - \frac{2GM}{c^2} r_c} = r_c \sqrt{1 - \frac{2GM}{c^2 r_c}}$$

$$r_{i\pm} = \frac{\left(r_c - \frac{GM}{c^2} \right) \pm r_c \sqrt{1 - \frac{2GM}{c^2 r_c}}}{2}$$

---

[142] Shirley MacLaine's New Age books "Out on a Limb", "Dancing in The Light". In this case Hal is "Dancing in The Dark" with the other Laputan Pundits not in The Light. ☺

Therefore, in Einstein's theory, the two isotropic roots are complex inside the singular event horizon when

$$r_c < \frac{2GM}{c^2} \tag{8.10}$$

The situation is even worse in Hal's PV theory. The exponential K is a nonanalytic Taylor power series as $r_i \to 0$.

$$e^{GM/c^2 r_i} = \sum_{n=0}^{\infty} \frac{1}{n!}\left(\frac{GM}{c^2 r_i}\right)^n \tag{8.11}$$

Therefore it has an infinite number of isotropic roots. Note that Hal's theory still obeys (8.7) just like Einstein's. It is easy to see that his infinite set of isotropic roots will also go complex when r_c starts increasing in the region of (8.11) as $r_i \to 0$.

$$\frac{GM}{c^2 r_i} > 1 \tag{8.12}$$

that Hal mistakenly thinks is the strong field limit.

Suppose the opposite that

$$\frac{GM}{c^2 r_i} \ll 1 \tag{8.13}$$

Truncate the series at n = 2 and we *almost* get Einstein's quadratic equation above for two isotropic roots for one curvature coordinate that is a control parameter. It is still a quadratic equation, but the coefficients are not exactly the same. That is for PV:

$$r_c = r_i e^{GM/c^2 r_i} \approx \frac{r_i}{0!}\left(\frac{GM}{c^2 r_i}\right)^0 + \frac{r_i}{1!}\left(\frac{GM}{c^2 r_i}\right)^1 + \frac{r_i}{2!}\left(\frac{GM}{c^2 r_i}\right)^2$$

$$r_c \approx r_i + \frac{GM}{c^2} + \frac{1}{2 r_i}\left(\frac{GM}{c^2}\right)^2$$

$$r_c r_i \approx r_i^2 + \frac{GM}{c^2} r_i + \frac{1}{2}\left(\frac{GM}{c^2}\right)^2 \tag{8.14}$$

$$r_i^2 + \left(\frac{GM}{c^2} - r_c\right) r_i + \frac{1}{2}\left(\frac{GM}{c^2}\right)^2 = 0$$

$$r_{i\pm} = \frac{\left(r_c - \dfrac{GM}{c^2}\right) \pm \sqrt{\left(\dfrac{GM}{c^2} - r_c\right)^2 - 2\left(\dfrac{GM}{c^2}\right)^2}}{2}$$

$$= \frac{\left(r_c - \dfrac{GM}{c^2}\right) \pm \sqrt{-\dfrac{2GMr_c}{c^2} + r_c^2 - \left(\dfrac{GM}{c^2}\right)^2}}{2}$$

$$r_{i\pm} = \frac{\left(r_c - \dfrac{GM}{c^2}\right) \pm r_c\sqrt{1 - \dfrac{2GM}{c^2 r_c} - \left(\dfrac{GM}{c^2 r_c}\right)^2}}{2}$$

$$r_{i\pm} \approx \frac{\left(r_c - \dfrac{GM}{c^2}\right) \pm r_c\sqrt{1 - \dfrac{2GM}{c^2 r_c}}}{2}$$

(8.15)

So the net PV result in this weak field limit is approximately the same as Einstein's except Einstein's result is not only for the weak field. We can see in the PV case as we truncate at higher and higher n we get more isotropic roots for a fixed r_c. I conjecture that as $r_c \to 0$ all the r_i are complex. It any case it is obvious that this non-covariant PV model is an ugly mess compared to Einstein's.

Therefore, Hal is in the absurd position that he has an infinite number complex radial isotropic coordinates in what he wants his strong field to be for a single fixed curvature radial coordinate r_c!

Finally, looking at his Tables I & II it is clear that Hal erroneously thinks his rest LNIF rods shrink locally isotropically and that the non-Euclidean radius R has $2\pi R >$ circumference C of a concentric circle is a global effect from integrating rather than a local differential effect. In fact, Hal's theory is spatially *locally* anisotropic just like Einstein's theory though with a different radial dependence. This is because his PV metric can be written as

$$ds^2 = \frac{1}{K}c^2 dt^2 - K dr_i^2 - r_c^2\left(d\theta^2 + \sin^2\theta d\phi^2\right)$$

(8.16)

Finally Hal offers no reason at all to think that K < 1 is good for propellantless propulsion. Indeed, IMO it is worse.

"You've Got Mail!"- Rants and Raves with Hal Puthoff

On 9/11/02 5:32 PM, "Puthoff@aol.com" <Puthoff@aol.com wrote:
In a message dated 9/11/02 5:22:35 PM, sarfatti@pacbell.net writes:
"The exponential solution is wrong it gives a singularity at space-like infinity. Hal and Ibison and Davies are completely confused on this..."

Hal Puthoff: Jack is the one who is completely confused about this, as the following demonstrates.

I wrote: This is a matter of physical interpretation of the formal symbols.

You have an infinite set of isotropic radial coordinate branches for only one curvature radial coordinate.

Furthermore, when radial coordinate $< GM/c^2$ your infinite set of isotropic radial coordinates go complex.

Einstein's GR in contrast has only two isotropic radial coordinates for each curvature radial coordinate. They go complex when radial coordinate $< 2GM/c^2$.

These two branches of the isotropic radial coordinate have physical meaning as different pieces of the curved manifold.

You claim EEP. That only works if PV is treated as a metric theory with geodesic structure. What is the meaning of your infinite set of isotropic radial coordinates? What is the Penrose diagram for this Monster? It is an interesting monstrosity and I continue to study it mathematically. Your PV solution is from the Island of Dr. Moreau. Dicke had no idea of Penrose diagrams and global methods when he came up with your mishugannah metric 40 + years ago. GR has come a long way since then Rip Van Winkle. It's time you awaken from your dogmatic slumbers.

Your claims are clearly absurd.

Hal wrote: Jack introduces an oft-used definition that size of area $4 \pi r_c^2$ (as used in standard Schwarzschild models) defines where one is.

I wrote: As I thought you do not even understand what your assistant Ibison showed. He correctly showed this result works not only for Einstein's GR but for Dicke's metric that you take as the corner stone of your new Temple of The Golden Calf.

Hal wrote: (The subscript stands for what one measures). If area is large, r is large, therefore one -> infinity; if area is small, r is small and one is closer to the mass. Simple, straightforward, seemingly unassailable.

I wrote: Yes unassailable in view of two embarrassing mathematical truths of your ridiculous proposal.

1. You have an infinite set of isotropic r for one unique curvature r.

Which one do you choose and why?

2. What does it mean when your infinite set of isotropic r go complex for real curvature r?

Your model here is so mathematically ugly, how you could wish to elevate isotropic r to a preferred status shows an egregious lack of theoretical insight IMO.

Hal wrote: However, in the PV modeling (with vacuum-distorted malleable rods)

I wrote: Balderdash as GR also has distorted rods. That is not an essential difference.

Hal wrote: as one approaches the mass it turns out that, indeed, the initial trend is for the measured area to shrink as in the Schwarzschild case. However, as one approaches the mass the *measured* area reaches a minimum and then begins to increase because the rods are beginning to shrink relatively rapidly compared to their length further out.

I wrote: I call this The Laputan solution. Again I ask.

Infinity of isotropic r for one unique curvature r — which one and why?

Complex isotropic r for real curvature r means what?

This same thing happens in GR — this boomerang and it means different regions of the curved manifold outside the horizon!

Hal, your topsy turvy Orwellian inversion is what I mean that you have been out so long it looks like in to you. ☺

Hal wrote: No problemo (except for Jack). He insists that the Schwarzschild modeling be overlaid on the PV modeling,

I wrote: That is a Red Herring Hal. You do not even understand what Ibison did! He showed that PV metric in rest LNIF of static source M is

$$ds^2 = \frac{1}{K}(cdt)^2 - Kdr_i^2 - r_i^2\left(d\theta^2 + \sin^2\theta d\phi^2\right)$$

Hal, you do not even understand the math of your own model!

Ibison showed

$$r_c = r_i\sqrt{K}$$

Hal wrote: and thus the increasing *measured* area (in PV) as one approaches the mass "must mean that one has turned around" and is heading back to space-like ∞ and therefore isn't approaching the mass" (Jack wrote this quote earlier that Hal cites)

Jack: Indeed, that's the well known boomerang effect in GR that leads to the two coordinate patches of the Penrose diagram outside the event horizon in the GR SSS vacuum Einstein bridge. The two isotropic r_i in the case of Einstein signal two parallel universes connected by a singular non-traversable wormhole! This is where Penrose's global topology of overlapping charts making an atlas to cover the curved manifold comes into its own. Similarly in your model where now you have an infinite set of r_i for only one r_c. What is worse for you is that your r_i go *unphysically* off the real axis into the complex plane in the real strong field limit, which is $r_c \to 0$. You obviously have not the slightest inkling of the great global revolution in GR post Dicke 1961.

Hal, you are like the Laputans Bunthorne[143] sings about:

"For art stopped short in the cultivated court of the Empress Josephine".

For the meaning doesn't matter for PV is only idle chatter of a transcendental kind .... ☺

Hal wrote: He can't think in terms of the PV modeling approach

I wrote: Correct, it requires a level of idiocy I have not yet descended to.

Hal wrote: whereby the area as measured by malleable rods can decrease then increase as one approaches the mass.

I wrote: The infinity and complexity of the isotropic r in the light of one unique real curvature r should give you pause in making such a proposal if you had any sense.

Now in fact your non-analytic monster metric has an infinity of parallel universes connect by some kind of topological Medusa of an infinitely branching wormhole that I am not mathematically equipped to decode as yet.

Hal wrote: PV may be incomplete; PV may be wrong; PV may be too "physical."

---

[143] "Patience", Gilbert & Sullivan parody on Oscar Wilde. It looks as though Puthoff's metric has only two sheets just like the Einstein-Rosen Bridge. However, finite Taylor series truncations to the exponential give more sheets, i.e., regions of positive real isotropic radial coordinate, like my Medusa image (added 10/21/02)

I wrote: Yes, yes and no. An infinity of complex isotropic r for one real curvature r Hal. What about that?

Hal wrote: Whatever.

I wrote: Were you a Valley Girl in one of your previous reincarnations? Next you will want to go shopping at The Mall on my credit card! ☺

Hal wrote: But PV is free to define its variables and their significance in whatever way it chooses.

I wrote: That is another predictable New Age "Create your own reality" remark. Did you get that listening to Gary Zukav on Oprah? ☺ Sure you can do that, but if you do then you cannot cite Lightman, Lee and Pope Dicke that your abortion of a theory obeys EEP. EEP assumes a metric theory with local covariance and geodesic structure and the Penrose global topology of manifold theory, which your Cargo Cult "toxic cosmology"[144] runs roughshod over like a wild boar in a china shop. You theory Hal is not even wrong not only for reasons you have not imagined, but probably also for reasons you cannot ever imagine. Can an old dog be taught new tricks? ☺

Hal wrote: What counts is what it predicts in terms of its defined variables and what the consequences are empirically. (Just as in GR generally, one can choose different coordinates to describe the same phenomenon - - physics does not depend on human choice of coordinate systems to describe it!)

I wrote: While we are on that subject. What do you claim to predict?

Hal wrote: So, on the above issue ("The exponential solution is wrong it gives a singularity at space-like infinity"), Jack's misstatement is a misstatement in logic;

I wrote: The only miss here is your pretense at a theory worthy of replacing Einstein's. It is clear you do not understand modern mathematical methods in GR post-1961 in particular global methods in manifold theory as started by Roger Penrose.

Hal wrote: it has nothing to do with math or physics. It has to do with not having the ability to think in terms of an alternatively defined system of coordinates (i.e., alternative to what he is familiar with).

I wrote: It has everything to do with math and physics and with my refusal to think like a New Age airhead. I have studied this ridiculous creature now for more than 25 years starting with est and Esalen. That Gary Zukav has made it to Oprah means this country is in deep doo doo in addition to Al Qaeda and Saddam, Hamas et-al. I am even more alarmed here because of Nick Herbert's support of Nazi apologist "historian", David Irving, since Nick for years was Dr. Physics at Esalen, the center of New Age Cargo Cultism. http://qedcorp.com/book/psi/hitweapon.html

Puthoff@aol.com wrote:

"In a message dated 9/28/02 11:18:29 AM, sarfatti@pacbell.net writes:

Yilmaz gets his exponential metric by Magick and it does not seem to jive with rest of his mumbo jumbo.

Hal wrote: "Yilmaz starts with a Lagrangian, derives Euler's Equations and solves them subject to the boundary conditions of a static spherical mass distribution. If that's mumbo jumbo, all of physics (including your own) is in trouble!

:-)

Collegially,

---

[144] Coined by Richard (Dick) Farley.

Hal

I wrote: Perhaps, I was only looking at first three pages of the paper Paul Zielinski gave me. Reading Yilmaz is a big job — Paul is working on it and will write a pedagogical paper.

Meantime it is obvious why Yilmaz[145], and you, are wrong on a deep physical basis, i.e. no way you can obey EEP.

Note 1. Why Yilmaz's physical idea is all wrong, i.e. in essential heuristic conflict with Einstein's equivalence principle no matter the fancy shmancy math.

On page 177 Pauli[146] writes of "final clarification ... Einstein proved .. the total energy and momentum of a closed system are, to a large extent, independent of the coordinate system, although the localization of the energy will ... be completely different for different coordinate systems ... one cannot assign any physical meaning to the values of the $t_{\mu\nu}$ [147]themselves, i.e. it is impossible to carry out a localization of the energy and momentum of the gravitational field in a generally covariant and physically satisfactory way. But the integral expressions (447) have a definite physical meaning." Pauli's classic 1921 Encyclopedia article modified in 1958 Dover

Remember when one writes

$$ds^2 = g_{\mu\nu}dx^\mu dx^\nu \qquad (8.17)$$

The coefficients $g_{\mu\nu}$ are *representations* [148]of an invariant form relative to a non-rotating local Cartan mobile frame whose subluminal moving center is on a time-like world line that is not generally geodesic. It's when that world line is not geodesic that we see gravity as locally equivalent to an inertial fictitious force[149] like the Coriolis force and the centrifugal force in rotating frames.

The reason is that the $t_{\mu\nu}$ must be non-covariant because what it really is, is the local frame dependent stress-energy density from near induction field electromagnetic reaction forces and fermionic quantum exchange pressures from Pauli's exclusion principle forcing the local frames off time-like geodesics[150] in Curve World. This is what happens when the elevator accelerates upward and you stand on a scale and your weight momentarily increases during the acceleration caused by electrical force etc. This $t_{\mu\nu}$ is zero if the elevator cable is cut, hence $t_{\mu\nu} = 0$. $t_{\mu\nu}$ is not a Diff(4) tensor in Curve World. It is only a Flat Land tensor that is *not* spin 2 gauge invariant.

The Flat Land Local Gravity Stress Energy Density imprint is simply $R_{\mu\nu}$ (Flat World)$_{NL}$ where in vacuum in Curve World

$R_{\mu\nu}$ (Curve World) $= R_{\mu\nu}$ (Flat Land)$_L$ $+ R_{\mu\nu}$ (Flat World)$_{NL}$

---

[145] "Toward a Field Theory of Gravitation", H. Yilmaz, Nuovo Cimento, 107B, N. 8, August 1992, p. 941
[146] Theory of Relativity, W. Pauli, Dover, 1958 paperback
[147] The *non-covariant* pure gravity stress-energy density pseudo-tensor under Diff(4) the Einstein gravity symmetry group.
[148] Mere shadows on the Wall of Plato's Cave in Book VII of The Republic.
[149] Obviously one feels these "fictitious inertial forces". They are measurable on a strain gauge. They are not delusory. They are physically real. "Fictitious" was a poor historical choice of word.
[150] The straightest possible slower than light world lines in curved space-time.

$R_{\mu\nu}$ (Flat World)$_{\text{NL}}$ = Sum of infinite series of Feynman scattering diagrams of the self-interacting massless spin 2 tensor field in Flat Land as, e.g. in Feynman's Cal Tech Lectures on quantum field theory of gravity in unstable Flat Land.[151]

$R_{\mu\nu}$ (Flat Land)$_{\text{L}}$ and $R_{\mu\nu}$ (Flat World)$_{\text{NL}}$ are separately global Flat Land Poincare group tensors, but they are not separately local Curve World Diff(4) tensors, though their SUM is! A similar thing happens in spin 1 QED as I recall in Feynman's papers. I will check that detail out anon.

This is the basic idea in Stanley Deser's paper on gravity and gauge invariance.[152] Spin 2 local gauge invariance in Flat Land induces Diff(4) local covariance in Curve World "After The Fall"[153], i.e. post translation subgroup of global Poincare group spontaneous symmetry breaking "More is different"[154] vacuum phase transition from unstable completely random high entropy micro-quantum zero point incoherent Flat Land to metastable emergent less random low entropy macro-quantum coherent Curve World with much smaller random zero point fluctuations.

This, I suspect, essentially solves the Λ cosmological constant problem, the Arrow of Time problem, the dark energy and dark matter problems, the source of 10^{20} ev superhigh energy cosmic rays and gamma sources, Arp's anomalous near quasar redshifts, and practical metric engineering of Star Gate Time Travel + free float global superluminal warp drive — all in one blow using only Einstein's geometrodynamics and traditional "More is different" emergent order physics well known from the many-body problem.

Practical metric engineering demands metastability of Curve World.

Eric Davis makes a completely unjustified claim in my opinion in his widely publicized MUFON 2001 paper on UFO physics:

http://198.63.56.18/pdf/davis_mufon2001.pdf

Eric Davis wrote: "Attached is my MUFON paper for all to read, again if necessary. You're stretching WAY beyond the facts here regarding what it is Hal and I have done and what it is I wrote."

Jack: I think not since this is what you did write on this:

"There is agreement between Puthoff's PV model and Yilmaz's approach in theoretical and experimental predictions for cases of interest in GTR and propulsion"

"The interesting feature of both the PV model and Yilmaz's approach is they both predict that the endpoint of (large-mass) stellar collapse is not a black hole, but is instead a "gray hole" possessing neither an event horizon nor a singularity."[155]

---

[151] "Feynman Lectures on Gravitation", Hatfield, Preskill, Thorne, Perseus, 1995

[152] "Self-Interaction and Gauge Invariance", GRG, 1, 1 (1970) pp 9-18

[153] A play by Arthur Miller about Marilyn Monroe if I remember correctly?

[154] Coined by P.W. Anderson for emergence of collective order from ground state symmetry breaking in the sense of the Goldstone theorem as in my UKAERE paper with Marshall Stoneham on the Jahn-Teller crystal lattice distortion in Proceedings of Physical Society of London, 1966-7 cited in AIP Resource Letter on Symmetry in Physics.

[155] I say this prediction is false. I say Puthoff and Davis do not know the difference between "strong field" and "weak field". They have been out so long it looks like in to them. (An allusion to Dick Farina's Cornell novel "Been Down So Long It Looks Like Up To Me" about the group I was part of in the Girl's Dorm Cornell Riots

"We are also exploring the application of the PV model to the Alcubierre warp metric, and expect K < 1 in this case since it is physically similar to traversable wormhole physics. However, we cannot predict at this time what potential alteration to wormhole and/or warp drive physics either the Yilmaz approach or the PV model will require.

Finally, Puthoff's PV model is the only alternative theory of gravity that has been successfully applied to explain the physical, anti-physical and physiological characteristics and performances of UFOs as described in section 3.2 (48). Puthoff showed that when the data from section 3.2 and reference 26 are taken together, these characteristics and performances can be reproduced by craft exploiting a technology that modifies the local space-time metric by varying K (K < 1, K > 1) as needed to generate the desired (propulsion and dynamic maneuvers and related) effects. It is possible that the PV model can provide either or both traversable wormhole and warp drive like manifestations within the context of UFO phenomenon." — Eric Davis

Jack's reaction: I see no evidence at all that K < 1 helps in anyway. Extraordinary claims such as Eric's above require extraordinary proof and you and Eric have given none! That is not the way physics should be promoted. Show what you have or retract!

Eric Davis wrote to me on 9/28/02:

"You are wrong with your claim (see Hal's repost of it below). My MUFON Symposium paper referenced both Yilmaz and Puthoff as two separate independent alternative GR theories, which yielded the same end-result for calculations they made to predict strong-field gravity physics."

I answered:

1. That's a delicate distinction. It it walks like a duck, if it talks like a duck .....

2. Hal's use of "strong field" looks like "weak field" to me. Both of youse guys with Yilmaz seem to have been out so long it looks like in to you.

What do you make of the fact that as implicit in Michael Ibison's[156] nice little PV formula

$$r_c = r_i e^{GM/c^2 r_i}$$

that for a given r_c, there are an infinity of r_i

Furthermore, that infinite set of r_i are unphysical complex numbers in the limit $r_c \to 0$, which is the TRUE "strong field" IMO. Your claim that $r_i \to 0$, all infinity of them, are strong field is bogus IMO. Indeed, since $r_i \to 0$ imply $r_c \to$ infinity this is weak field obviously — in PV terms.

Hal Puthoff wrote on Oct 4, 2002:

"I repeat:

---

at President Deane Mallot's house.). Puthoff needs an infinite number of isotropic radial coordinates for each unique curvature radial coordinate. Furthermore, the real strong field is when the curvature radial coordinate -> 0. The infinity of isotropic radial coordinates become complex numbers in that limit showing the existence of an event horizon for the exponential isotropic metric. This vacuum space time has a pathological infinity of non-traversable wormhole mouths each giving a chart on the Penrose diagram. This monster can be likened to Medusa's hair of writhing snakes.

[156] Dr. Ibison worked with Basil Hiley at Birkbeck College, University of London, and now works for Hal Puthoff in Austin, Texas. Hiley was David Bohm's assistant for 30 years. I was an Honorary Research Fellow at Birkbeck with Bohm in 1971 and I catalyzed the Uri Geller tests at Birkbeck in 1973 that Martin Gardner describes in "Magic and Paraphysics" in "Science, Good, Bad and Bogus".

<< Now it's your turn. Answer his specific charge as to why his (Ibison's) correction to your faulty critique is not correct and telling. >>

Ibison is an excellent theoretical physicist. You need to come up to his standard and answer his correction.

I replied on Oct 4, 2002:

Sorry Hal but you are acting obtusely as anyone who reads what is actually there can see.

First of all Ibison admitted to me that he is an amateur in general relativity. His primary field is quantum mechanics under Basil Hiley at Birkbeck if I am not mistaken?

Second, there is nothing I say that contradicts Ibison's math! I have no quibble with his algebra on this very tiny detail. Indeed, I use Ibison's nice little formula for your PV! This is high school algebra that even a practical metric engineer should be able to understand. You don't have to be a rocket scientist to get this.

Third, the more substantial issue is the informal one of interpretation of radial curvature coordinate r_c and isotropic radial coordinate r_i in general independent of the detailed dynamical action for a given model, i.e. GR vs PV. This depends on manifold theory and global topology of manifold which neither you nor Ibison know diddly squat about.

You are not free to make up your own rules on that like you do.

In this case, your attempt to replace curvature r_c by isotropic r_i as the true measure of strong vs weak field is physically untenable and downright stupid if you understand anything at all about Penrose diagrams.

First of all your common sense should tell you something is amiss if you get a multiple set of isotropic r_i for a single curvature r_c.

Second, it's even more serious when your isotropic r_i become *complex numbers* in precisely the "strong field" region used by Einstein in GR.

Now Einstein's GR has *two* isotropic r_i for one curvature r_c. These two isotropic r_i are complex inside the event horizon for the SSS vacuum solution.[157]

This is indeed why the two isotropic r_i are two coordinate patches outside the event horizon forming an Einstein-Rosen bridge or non-traversable wormhole in the global topology of the manifold for Einstein's SSS vacuum solution.

Now in your model using Ibison's nice little PV equation relating isotropic r to curvature r you have in your PV an infinity of isotropic r_i for a single curvature r_c. Like GR your infinity of isotropic r_i also go complex when curvature $r_c \rightarrow 0$ in the traditional strong field limit. This should give you pause, but it obviously doesn't.

Each isotropic r_i must be a coordinate patch and your non-analytic $y = e^{\wedge}1/x$ is the source of your pathological infinity, your excess mathematical baggage of an infinity of isotropic r_i. Hence I call your monster solution the Medusa metric. This is an interesting example of pathological physics.[158]

Hal you are simply not connecting the dots here. You have never addressed this and neither has Ibison.

"Anything Goes" is a good musical but not a good approach to theoretical physics.

Neither is "Create your own reality" in spite of Niels Bohr.

---

[157] Static Spherically Symmetric Schwarzschild vacuum solution, e.g. "Lorentizian Wormholes" by Matt Visser, AIP Press.

[158] A good example of "nonsense physics" that is in the Einhorn murder trial in which "viruses beamed on microwaves" like Laputan "sunbeams from cucumbers" is mentioned as a "defense".

Eric Davis wrote on 9/28/02: "I never addressed the validity of Hal's theory in terms of Yilmaz's theory. The validity of Hal's theory is not in terms of anything Yilmaz did. The validity of Hal's model is based on its ability to make predictions that keenly match observed phenomena. And THAT is the crucial scientific method of validation."

I wrote: I agree 100%. So where is the beef? Show us all right now how Hal using PV has the "ability to make predictions that keenly match observed phenomena". I note your use of the word "keenly" and I call your bluff. Since we are talking about the alleged phenomena in your MUFON 2001 paper simply show us. You have a nice little list there of UFO phenomena. Go down your list shown below and check off how Hal's PV "keenly" mind you "match observed phenomena". Since you said you once saw Hal do this at NIDS what's the problem? I am reminded of H... K... at ISSO who kept boasting he had the solution to vacuum propulsion and release of vacuum energy, but never allowed any one who could check his claims to see what he really had in terms of mathematical theory. Do you also claim that The Men In Black Ops will kill us all if you reveal this information?

So simply put up or shut up. If you can show by correct examples that Hal's PV has the "ability to make predictions that keenly match observed phenomena" then you will win the debate quite obviously. Extraordinary claims such as yours and Hal's require extraordinary proof. So far no proof at all has been forth coming. Simply astound us all! Well?

Eric Davis's List of UFO Phenomena

"3.0 UFO Phenomenology from http://198.63.56.18/pdf/davis_mufon2001.pdf
Paul Hill[159] has already delineated and characterized UFO performances and dynamics in his excellent book (26). From a rigorous aeronautical and physics analysis of many cases (the unexplainable, non-prosaic ones), Hill concluded that UFOs are craft that would have to

---

[159] A very good USG aeronautical engineer WWII vintage who wrote the only good technical book on UFO physics "Unconventional Flying Objects". I think Eric Davis's list here is quite interesting. My disagreement is not with his empirical list but with his claim that Puthoff's PV model and/or Yilmaz's EEP-violating model has any relevance whatsoever to explaining alleged flying saucer flight dynamics.

utilize an engineered 'acceleration-field technology'[160] in order to manifest their various performance characteristics. "Acceleration-field"is the old fashioned term for space-time metric. Wormhole-stargates and the Alcubierre warp drive metric (27) are examples of modern space-time metric engineering concepts, both of which require engineering of the vacuum to mine the negative energies needed to generate such metric modifications.

Jacques Vallee has also analyzed UFO cases over four decades and summarized his findings in several excellent, groundbreaking books and articles (28,29,30,31 - the key references). Vallee concluded that UFO phenomenon is consistent with a technology centered on a craft using a very revolutionary propulsion system, which possesses an anti-physical dimension in addition to others. The phenomenon is the product of a technology in the sense that it is a real, physical, material object. The physical characteristics of UFOs is as follows (adapted from 28,29,32 and the NIDS database):

Ø witnesses describe an object that occupies a position in space*

Ø moves as time passes*

Ø interacts with the environment through thermal effects as well as light absorption and emission*

Ø produces turbulence*

Ø when landed, leaves indentations and burns from which approximate mass/energy figures can be derived*

Ø gives rise to photographic images*

Ø gives rise to electric, magnetic and gravitational disturbances*

But UFOs also manifest anti-physical effects by using advanced physical principles. These anti-physical effects are as follows (adapted from 28,29 and the NIDS database):

Ø sinking into the ground*

Ø shrink in size, grow larger, or change shape*

Ø becoming fuzzy and transparent on the spot*

Ø divide into two or more craft, several of them merge into one object at slow speed*

Ø disappearing at one point and appearing elsewhere instantaneously*

Ø remaining observable visually while not detected by radar*

Ø missing time/time dilation*

Ø topological inversion/space dilation (UFO was estimated to be of small exterior size/volume, but witness(s) saw a huge interior many times the exterior size)*

Ø balls of colored, intensely bright light under intelligent control*

Ø Doppler blueshifting and redshifting effects of moving and motionless UFOs*

The physiological reactions caused by UFOs are (adapted from 28,29,32 and the NIDS database):

Ø burns**

Ø sounds (beeping, buzzing, humming, sharp/piercing whistling, swooshing/air rushing, loud/deafening roaring, sound of a storm, etc.)**

Ø vibrations*

Ø partial paralysis**

Ø extreme heat or cold sensation*

Ø odors (powerful, sweet or strange fragrance, rotten eggs, sulphurous, pungent, stinking, musky-like, etc.)**

---

[160] Jack note: I claim that my new macro-quantum vacuum Λ field physics that is completely consistent with Einstein's gravity physics explains Paul Hill's "acceleration-field technology".

Ø metallic taste**
Ø pricklings**
Ø temporary blindness when exposed to the objects□ light*
Ø nausea**
Ø bloody nose and/or ears; severe headache**
Ø difficulty in breathing**
Ø loss of volition**
Ø drowsiness in the days following a close encounter**

There are psychic effects triggered by UFOs either purposely or as a side effect of the presence of the UFO. These are (adapted from 28,29 and the NIDS database):

Ø impressions of communication w/o direct sensory channel Ø levitation of the witness or of objects and animals in the vicinity*

Ø poltergeist phenomena: motions and sounds w/o a specific cause, outside of the observed presence of a UFO Ø maneuvers of a UFO appearing to anticipate the witness□ thoughts Ø premonitory dreams or visions Ø personality changes promoting unusual abilities in the witness Ø healing

3.1 Example Cases from the NIDS Database

UFO witness descriptions are the database presently available for examining the wormhole hypothesis along with the meager physical data acquired by surveillance equipment (see for example, references 36-39). And we recognize that witness reports are not rigorous from the standpoint of collecting physics data. Of the more than 650 cases investigated by NIDS, several dozen clearly portend wormhole manifestations. Particular examples include field research NIDS conducted in northeastern Utah whereby the following example data was acquired:

Ø intensely bright, colored balls of light under intelligent control; either monochromatic or changing color; possessing either smooth or variable liquid turbulence-like surface/internal texture; maneuvering/hovering near people and around property; brightening or fading and blue/red Doppler shifting when appearing or disappearing

Ø very large, very bright orange-colored opening in the daytime sky; a completely different or foreign looking sky was seen through the opening; an object was seen (through rifle spotting scope) moving through the opening at rapid speed

Ø faint light appears in the air a few feet above a dirt road; light grows in intensity becoming very bright; bright light then becomes a hole that opens up (growing from 1 to 3 feet diameter) and from within which another light is emanating; a large, black creature (~ 400 lbs., 8 to 9 feet tall) is seen crawling out of the hole (as seen through 3rd generation military night vision, hole appeared 3-dimensional with tunnel-like interior), it stood up and ran away into the surrounding dark of night; the brightly lit hole closed and faded away" by Eric Davis

More complete PV theory remarks

Eric Davis continued in http://198.63.56.18/pdf/davis_mufon2001.pdf. Puthoff's key PV paper is at http://xxx.lanl.gov/abs/gr-qc/9909037

"Puthoff's approach, known as the polarizable-vacuum (PV) representation of general relativity, treats the vacuum as a polarizable medium (42). The PV approach treats space-time metric changes in terms of equivalent changes in the permittivity and permeability constants of the vacuum, $\varepsilon_0 \& \mu_0$, essentially along the lines of the $TH\varepsilon_0\mu_0$ methodology used in comparative studies of gravitational theories (see references cited in 42). Such an approach, relying as it does on parameters familiar to engineers, can be considered a 'metric engineering' approach. Maxwell's equations in curved space are treated in the isomorphism of a polarizable medium of variable refractive index in flat space (see references cited in 42); the bending of a light ray near a massive body is modeled as due to an induced spatial variation in the refractive index of the vacuum near the body; the reduction in the velocity of light in a gravitational potential is represented by an effective increase in the refractive index of the vacuum, and so forth. As elaborated in reference 42 and the references therein, PV modeling can be carried out in a self-consistent way so as to reproduce to appropriate order both the equations of GTR, and the match to the classical experimental (PPN parameters and other) tests of those equations. There is agreement between Puthoff's PV model and Yilmaz's approach in theoretical and experimental predictions for cases of interest in GTR and propulsion.

Specifically, the PV approach treats such measures as the speed of light, the length of rulers (atomic bond lengths), the frequency of clocks, particle masses, and so forth, in terms of a variable vacuum dielectric constant K in which vacuum permittivity ε_0 transforms to $\varepsilon_0 K$, vacuum permeability to $\mu_0 K$. In a planetary or solar gravitational potential K > 1 while K = 1 in 'empty' remote space. In the former case, the speed of light is reduced, light emitted from an atom is redshifted as compared with a remote static atom (K = 1), clocks run slower, objects/rulers shrink, etc.[161]

The interesting feature of both the PV model and Yilmaz's approach is they both predict that the endpoint of (large-mass) stellar collapse is not a black hole, but is instead a gray hole possessing neither an event horizon nor a singularity. Such a body would simply be an extremely collapsed state of matter. Recent astronomical observations have reported that neutron stars more massive than the lower limit collapse mass for black holes have been discovered, which severely contradicts the strict mass constraints placed on neutron star formation by Einstein's GTR (46). And it still has not been experimentally possible to positively confirm black hole candidates on the basis of predicted strong gravitational field effects occurring outside their alleged event horizons. No astronomy experiment has positively observed and measured a black hole's event horizon. And it is impossible to experimentally confirm the existence of black hole singularities given the inaccessibility of the black hole's interior to observation and measurement as predicted by Einstein's GTR. Einstein's GTR is not a flat-out failure; it is just in need of some repair.

Under certain conditions the space-time metric can in principle be modified to reduce the value of K to below unity thus allowing for faster-than-light (FTL) motion to be physically realized. In this case, the local speed of light (as measured by remote static observers) is increased, light emitted from an atom is blue shifted as compared with a remote static atom, objects/rulers expand, clocks run faster, etc. In fact, Puthoff has analyzed certain special black hole metrics and found K < 1 from his model. We are both examining whether the traversable wormhole metric will also lead to K < 1 within the PV model (46). In fact, we have reason to believe that there will be such a solution on the basis that traversable wormhole metrics are an exact metric solution to Einstein's GTR such that they do not possess physically/mathematically pathological features such as an event horizon or singularity. This has been supported by exact high-order nonlinear quantum electrodynamic

---

[161] Jack note: Puthoff's Table's I and II give a prediction of local spatially isotropic ruler distortion that contradicts Einstein's local spatial anisotropic ruler distortion. Therefore Puthoff's claim that his PV theory agrees with Einstein in the weak field limit is false.

analysis of the vacuum cavity within a Casimir-effect capacitor experiment showing that the speed of light will in fact increase, as the Casimir-effect energy grows more negative within the cavity (47). This is what we would expect when a traversable wormhole effect manifests itself within a region of squeezed vacuum (recall section 2.6). We are also exploring the application of the PV model to the Alcubierre warp metric, and expect K < 1 in this case since it is physically similar to traversable wormhole physics. However, we cannot predict at this time what potential alteration to wormhole and/or warp drive physics either the Yilmaz approach or the PV model will require." – Eric Davis

Time Travel to the Past?

Stephen Hawking poses the "chronology protection conjecture". Very incomplete not quite quantum gravity calculations suggest that as soon as time travel to the past through a traversable wormhole Star Gate is about to click in, there is an infinite blue shift of light through the wannabe time machine that fries everything to a crisp. However, in my new theory we should not directly quantize the classical geometrodynamic Einstein field at all. This geometrodynamic field for Curve World is an emergent broken symmetry collective macro-quantum "More is different" mode of the quantum gauge source and force fields in the unstable Dirac-Fermi Sea micro-quantum globally flat false vacuum that disappears in the phase transition from Flat World to Curve World. Therefore, it's a new ball game for time travel to the past as in the "Satori Trilogy" incident ~ 1980 and the "Spectra Contact" in 1953. The Fat Lady has not sung on this issue. Indeed there is empirical folklore that time travelers from our future have been here and are here affecting our Destiny Matrix. Indeed, I may have been in contact with them. I cannot prove that of course unless this new physics of mine turns out to be correct. Too soon to say. Definitely the dark energy $\Lambda > 0$ is what we need to make Star Gates even if we cannot use them to time travel to the past. My bet is that we can time travel to the past and will![162]

The Ira Einhorn Holly Maddux Murder Mystery

Ira Einhorn was the book agent for the original Space-Time and Beyond published by Dutton in 1975. Bob Toben did the cartoons and I wrote 95% of the original text.

"From: Annika Einhorn
Subject: A plea from Annika Einhorn for character witnesses at Ira's trial
Date: Sunday, September 29, 2002 2:47 AM

The trial for Ira starts tomorrow Monday. I wanted to share that there has been some very interesting developments. We have found a fourth person who saw Holly alive after she was supposedly dead in the trunk, according to the prosecution. This witness is a retired policeman who actually knew Holly and Ira as he was stationed in their area during the

---

[162] We cannot use a traversable wormhole Star Gate to travel to a time before the Star Gate was built. However, there may be very old Star Gates around from ancient extra-terrestrial advanced civilizations in this universe and from the parallel brane universes next door in Super Cosmos across a thin hyperspace "weak link" in still another potential application of the Josephson Effect?

Move uproar. One other person who saw her is dead, but the two others have been found and will testify.

Then there is the forensic issue, which simply does not correspond to the prosecution's case, there was no blood or no human protein in the apartment where as if she would have been killed there one should have found plenty of evidence.

Another issue is Joyce and Cindy. The two women that supposedly was asked to move the trunk by Ira and who felt such a revolting smell that they withdraw. It was Joyce's testimony that got Ira convicted more than anything else in 93.

Cindy will now testify saying that she remembers nothing of a trunk. She will also testify to that Joyce told her she called Ira in 97. A phone call I remember very well indeed as it woke us at 6 in the morning. In it she apologized to Ira for having lied at the trial but she said the DA pushed her to, and she wanted to help Holly, and she did not realize that it would turn against Ira. She was very upset and agonized.

I am not trying to convince you of Ira's innocence, even though I do believe he is indeed innocent, but I am trying to activate you and possibly some others to come forth and talk about your memories of Ira of the time, not as he has been painted by Levy and the DA and the press; leaving the judgment of guilt or innocence to the jury. You, his old friends, who speak out will at least give him a fairer chance as today he is very much lacking support of character witnesses. He needs seven.

A wonderfully supportive testimony would have been given by Prof. Stafford Beer, but unfortunately he has been very sick and died on September 23rd. Stafford tried to get a different message through to Steven Levy[163] for the book, but Levy chose to exclude it.

Please send this plea on to anyone you can think of that could possibly step forward.

The person to contact is Richard Strohm.

He is a retired Philadelphia Homicide Lieutenant working on Ira's case.

Richard T. Strohm & Associates

2509 S. Broad St. - Suite 203

Philadelphia, PA 19148

215-468-9969

email RTSData@aol.com

I am doing direct TV tomorrow, Monday the 30th of September, the opening of the trial. The Early Show with CBS in the morning. Larry King Live with CNN in the evening. Bill Cannon, Ira's lawyer, is doing Connie Chung with ABC in the evening.

Warm regards, and a heartfelt thanks to anyone who has the courage to stay true to their positive memories of Ira.

Annika Einhorn

note: There is an article in today's Sunday Philly Inquirer interviewing Annika, "Einhorn's #1 supporter" posted at

http://www.philly.com/mld/philly/news/4170907.htm"

The retired policeman's testimony is key for establishing reasonable doubt as is the alleged lack of blood and protein in the apartment. Ira was a complete slob and could not have cleaned his own apartment, which I did see in 1974 with Bob Toben and Sharon Moore as we were writing Space-Time and Beyond.

---

[163] "The Unicorn's Secret". Levy interviewed me for the book.

Speculative wild half-baked testimony from people like me who knew him mainly by correspondence would not help his case and could indeed hurt it as people like me would seem as nutty as John Nash in "A Beautiful Mind" to The Jury.

Ira will have to serve time for jumping bail even if he beats the murder charge. He would have beaten it for sure had he not run away in the first place. His case was then much stronger than O J Simpson's. There is a parallel here with Congressman Gary Condit who was apparently falsely[164] effectively convicted by the media for the murder of Chandra Levy.

Jack Sarfatti's reply to David Crockett's rant below:

Memorandum For The Record on Bogus Claims of Anti-Gravity Free Energy

I know as a Ph.D. physicist that essentially none of the information on free energy and anti-gravity that Ira had access to in the 70's was real.[165] It was all bogus Cargo Cult pseudo-physics. Ira was no physicist and he could not tell the difference anymore than my young gullible friend, Joe Firmage, can today. Ira was basically a useful idiot in humint jargon used to spread disinformation and misinformation on these topics for covert Cold War Psy Ops on both sides of the Iron Curtain. The same game is going on today with the UFO Disclosure Movement, only some of the players have changed.[166] Little did the Puppet Masters back then realize that some real gold lay under the pile of "macroshift" the Fool's Gold. Indeed, they got more than they bargained for it now appears.

Holly may have been murdered by the KGB, the Serbians, or by the Iranians or by the American Nazi Party or by person's unknown, or even perhaps by Ira. I simply do not know. I do not believe any agency of USG Federales murdered Holly. Holly may have died from her diabetes forgetting to take her insulin. Her body could then have been mutilated and planted in Ira's storage room outside his apartment. She may have been abducted when she left Ira's apartment to go to the store. A clever frame would use Ira's past patterns with women to frame him - to make it convincing.

Had Ira not run away he would have walked because he was deep enough into Black Ops and had enough powerful enemies who had opportunity, motive and lack of scruples to frame him.

None of the physics Ira was working on with Colonel Tom Bearden had any chance of ever working. It was Bogus Cargo Cult Pseudo Physics as are all the current claims by UFO Disclosure, by Ban US Space Weapons Movement, and by Nick Cook in the bogus "Hunt for the Zero Point" This book is a black mark on Jane's Defense Weekly IMO.

The only real physics here is recent in
http://stardrive.org/Jack/PT81502.pdf
http://stardrive.org/Jack/nextforce.pdf

Now Ira was promoting my work at the time. I was also, perhaps inadvertently, perhaps advertently, derailed at Esalen in Big Sur by George Koopman's addiction at approximately the same time that Holly disappeared ~ 1977. This lead to a bruhaha between me and the

---

[164] Article in San Francisco Chronicle as this book goes to press about a defective lie detector test on a convict who assaulted two other women in the same park where poor Chandra's remains were found.

[165] Nor is it real today despite unreliable books like Nick Cook's "Hunt For The Zero Point".

[166] Some of the old ones are still around http://qedcorp.com/book/psi/hitweapon.html

Werner Erhard est organization that Stephen Schwartz says was a Cold War Soviet KGB Front. So these connections are perhaps relevant I do not know.

It is conceivable that I would have made the UFO physics breakthrough of 2002 earlier as Ira would have probably gotten me a lot of money from his rich patrons like The Bronfman's et-al. However, my 2002 discovery of the Λ zero point field from the macro-quantum vacuum Bose-Einstein condensate only happened because of the very recent empirical discoveries of dark energy accelerating universe and the enormous amount of dark matter and the surprisingly small amount of star stuff in the universe only recently understood. Crockett's allusions to "the new-energy technologies to replace fossil fuel power" is Cargo Cult Bogus. Crockett is a New Age Airhead Crackpot on these issues.

Ira was working with Jacques Vallee on ARPA pre-Internet issues that is true.

Gene Roddenberry's Star Trek only happened because of his participation with Andrija Puharich's circle that Ira and I were part of.

It is curious that my physics of Making Star Trek Real has now reached critical mass.

Ira's involvement with the Tesla Archive may be relevant because of Jim Corum's similar involvement. Corum is funded now by Senator Robert Byrd at the ISR Think Tank in Senator Byrd's West Virginia home county. My own work increasingly points in the Tesla direction though not in the way that New Age Flakes promote it.

Davy Crockett's Rant

"The Philadelphia "Unicorn Killer" retrial, for a murder that compares gruesomely to the O J Simpson case, has related issues that are much more far-reaching — especially if Ira Einhorn's claims are true that he was framed by domestic or foreign intelligence agencies to silence his multi-issue activism in the 60's and 70's — and especially as the US prepares again to go to war to protect American oil supplies while suppressing the new-energy technologies to replace fossil fuel power that Einhorn might otherwise have helped implement by now. The retrial is expected to last a month. Will its expected broad media coverage help or hinder exposure of critical related issues?

Will his wife, Annika Einhorn, be able to mention any of the information below from Ira which substantiates his claims that he was framed for murder to shut him up? From France she will be on the CBS's Early Show tomorrow morning and CNN's Larry King Live in the evening. Bill Cannon, Ira's lawyer, is doing Connie Chung with ABC in the evening.

Below is a compilation offering reasons why Ira Einhorn's message may be too dangerous for his own good.

More info on these new-energy technologies, articles, links, are at
http://groups.yahoo.com/group/new-energy-solutions"

Jack's comment. I do not think any of the information on the above link can be trusted. It should be treated as disinformation and misinformation IMO[167].

"Subject: text, Ira Einhorn's "A Snapshot of my 70's", with footnoted documents
Date: Sunday, September 29, 2002 11:31 AM
[This post at http://groups.yahoo.com/group/ira-einhorn/message/211

---

[167] In my opinion

has an attached 4pp efax file containing a 1pg letter from Alvin Toffler, author of "Future Shock", and 3pp copy of Winter '77-'78 Co-Evolution Quarterly article containing additional info footnoted to bracketed numerals added to following text]

————begin text transcription of faxed article—-

A Snapshot of my 70's by Ira Einhorn

September 1, 2002

[transcribed to text, lightly edited for spelling and punctuation only; handwritten 9pp original as .jpg files at http://www.sonic.net/~west/ira]

"Stress is information that the body is unable to accomodate; pollution is material that the environment is unable to recycle within human time frames."

"Nanoseconds now can the emotions follow?"

—- Both quotes Ira Einhorn, 1970

"In the mid to late 60's the accelerating destruction of the planetary interconnecting web, which is our life support system and without which we cannot survive, became an emergent possibility to a few of us activists who took the trouble to read the then available ecological literature. We intuitively sensed the need to open a new front in the "movement" battle, for Chicago '68 was already pointing towards Kent State and the violence of frustration that lead to the Weathermen and other similarly doomed and fragmented groups.

Earth Day in Philadelphia was conceived as a necessary partnership between business, academic, political and activist interests, for environmental protection requires a conscious restructuring of all we do.

All of my subsequent work in the business community, as social change agent, consultant, futurist and learner, came out of that initial partnership brokered by Philadelphia Chamber of Commerce president Thatcher Longstreth.

I did not give up my participation in opposition to the war in Vietnam, my work with the black organizations that still allowed white participations, my campaigning for a sensible drug policy, my spreading of information about CIA involvement in the heroin trade, my investigative work on the assassination of J.F.K., my study of various techniques for changing consciousness, or my work in futurism, but rather shifted my focus and added a new arrow to my quiver.

It soon became obvious to me, after an enormous amount of reading and intense dialogue with many scholars, who knew much more than I about the coming ecological disaster, that ecocide was our species' future, if we did not find a way to bring about massive change.

The cosmetics of "greenwash" would buy time, but I knew from a deep study and intense listening to the marvelous lunchtime raps of Ian McHarg that even his comprehensive ecological planning structures, which were light years ahead of the brilliant work of Amory Lovins and so many others, would not halt our lemming like march towards an ecological abyss.

Hence my shift towards activities during the early 70's that were incomprehensible to my most radical eco-friends like ex-digger Peter Berg.

I felt that the entire planet needed the kind of shock that Arthur C. Clarke provides at the conclusion of "Childhood's End," plus a new kind of technology that would allow us to live without destroying the delicate balance that supports us. I later realized, due to my intense editing of many versions of Arthur Young's two major texts, "The Geometry of Meaning" and "The Reflexive Universe", that we also needed a comprehensive metaphysic that would subsume science and fold it into a more inclusive value laden structure.[168]

My study of Brecht, Artaud and Gurdjieff taught me much about shock and its effects. My study of UFO's quickly made me aware that the coverup of such information, irrespective of its meaning and explanation, was a "CosmicWatergate,"[169] but the elusive pattern of the recurring flaps made me aware that the phenomenon was beyond immediate use, though I sensed that its attendant psychic phenomena were of great possible use.

Thus, when the opportunity to work on the Uri Geller project with Andrija Puharich emerged in 1971, as a result of my convincing my editor/friend Bill Whitehead[170] to republish two of Andrija's earlier books, for one of which, "Beyond Telepathy", I wrote an introduction, I quickly combined my newly developed corporate contacts, Andrija's networks and those past "movement" friends able to deal with "magic," into my Network called by some "the internet before the internet existed."

The Network was originally set up under Bell of PA, then the local telephone company, with the approval of the then president Bill Cashel who later became the vice-chairman of A.T.&T.[171], at that time the largest corporation in the world, to service the information needs of those people linked with Andrija and I in the Geller Project. That Network grew during the 70's to over 300 key people in a multitude of human disciplines in 26 countries on both sides of the Iron Curtain. Each piece of information was circulated separately to those people in the Network that I thought would be interested in the information. Bell maintained a card file of names and addresses and did all the duplicating and mailing. No money ever changed hands. It was a barter arrangement between myself and the world's largest corporate entity. The information circulated on the Network soon grew to encompass emerging information in a large number of fields. The AT&T "far watchers" – those paid to eliminate surprises — soon came to talk to me at regular intervals, host me at AT&T headquarters and take me to Bell Labs, for I was circulating both articles and books that they should have known about but didn't.

What Geller could do, I saw a lot of it first hand, indicated that the basic physical framework in which physics operated was inadequate and that so called "free energy" devices — devices that would solve our energy problem and end what is now called global warming and allow for the decentralization of most economic activities — could become a reality.[172]

Hence I circulated all previously known anti-gravity information and all the emerging work on "free energy" devices.[173]

---

[168] Jack's comment: Arthur was a rich powerful man in our group, but I never thought much of his ideas about physics.

[169] Jack's comment: Dan Smith, a wealthy Blue Blood "UFO Aviary"eccentric like Arthur Young has a similar idea that he calls "The Eschaton". Dan says reading my original Space-Time and Beyond of 1975 blew his mind and changed his life. It was an epiphany for Dan. http://stardrive.org/cartoon/dan.html

[170] Jack's comment: edited original Space-Time and Beyond at Dutton that I wrote most of.

[171] Jack's comment: A mentor of mine, Marshall Naify wound up owning a large chunk of ATT later on.

[172] Jack's comment: Ira was, and is, deluded about free energy devices. They simply do not exist. We checked out many of these New Age claims at ISSO in 1999-2000. They were all bogus and silly.

[173] Jack's comment: For the record I know as a Ph.D. physicist that none of the information on free energy and anti-gravity that Ira had access to in the 70's was real. It was all bogus Cargo Cult pseudo-physics. Ira was no physicist and he could not tell the difference anymore than Joe Firmage can today. Ira was basically a "useful

Unfortunately, all new technology can be used as weaponry as well as for human benefit. So, I was soon up to my ears in a multi-pronged intelligence game that is still waiting to be unraveled.

A small subset of the Network soon found itself monitoring the "Russian Woodpecker" — a signal emanating from Soviet territory that appeared to have mind control properties[174], according to Eldon Byrd, the creator of a mind control device for the Navy, the CIA admitted as much in 1986, a decade later. Gathering together information from many sources, I published the first public article on the issue in the "Co-Evolution Quarterly"[175] in 1975. [1]

I was then inundated with mail from all over the world. One of my correspondents was P.K. Dick, [http://www.disinfo.com/pages/article/id773/pg1/] probably our best science fiction writer, whose work is slowly bleeding into popular consciousness through the movies: Ridley Scott's "Alien" and Spielberg's latest with Tom Cruise in the lead, among others, are based upon his prescient work. Dick wrote me a series of letters about mind control, that I circulated to a small group, and then poured out 500,000 words about the issues we discussed.

I also began receiving badly translated reports from all over the Soviet Union of psychotronic/mind-control weaponry. Weaponry so chilling that I only shared some of the content, not the actual reports, with two people: Arthur Koestler[176] and Stafford Beer. (see his letter to my original attorney: Norris Gelman). [2]

I also circulated a general letter to the entire Network about mind control issues that Alvin Toffler, author of "Future Shock", and I wrote together. [3]

I sat on all of these reports for more than a year and a half. I spoke to Arthur and then Stafford about the issue only after I was warned by Tony Judge and others that they felt my life was under threat.

At the time of my arrest, I had received over 200 reports, mailed from all over the world, but obviously originating from behind the Iron Curtain. In the summer of 1977, I planned a fact finding trip that would have taken me to many sites behind the Iron Curtain. In the Spring of 1977, the CIA contacted me through an intermediary, but I declined to cooperate under the circumstances. I wanted direct contact. My Iron Curtain part of the trip was cancelled after Los Angeles Times reporter Robert Toth was arrested in Moscow, supposedly in possession of material that was similar to that for which I was looking.

While this part of my life was going on, I was circulating edge information all over the world, running physics and consciousness weekends in a number of places, giving innumerable lectures and working on many other projects including being the director and money raiser for Sun Day in Philadelphia which saw our symbol adorn the top of the electric company building. I did 42 radio and television shows during the 14 day run up to this 2 week long event. At that time I was also involved in two major futures studies: one involving the future of communication for the Canadian Telephone Company, under the

idiot" in humint jargon used to spread disinformation and misinformation on these topics for covert Cold War Psy Ops on both sides of the Iron Curtain. The same game is going on today with UFO Disclosure, only some of the players have changed. Little did the Puppet Masters back then realize that some real gold lay under the pile of "macroshift" the Fool's Gold. Indeed, they got more than they bargained for it now appears.

[174] See Martin Gardner's "Magic and Paraphysics" in "Science, Good, Bad and Bogus"

[175] Published by Stewart Brand an advisor to Jerry Brown.

[176] Jack's comment: I was with Arthur Koestler and Arthur C Clarke at the Uri Geller tests at Birkbeck College with David Bohm and John Hasted that I initially arranged with Brendan O Regan.

auspices of Jacques Vallee, using his newly developed computer conferencing technique; the other a study I conducted myself for a small local multi-national corporation. In addition, I was constantly finding books for my editor Bill Whitehead to publish and beginning to lecture on the principles behind Networking for major corporations.

To substantiate the reports about the mind control properties of the "Russian Woodpecker," Andrija Puharich built a smaller version of the technology and tested it on some of our core group in private and in a number of public spaces — indicating a devastating ability to modulate human behavior.

So when the opportunity arose, after a series of dinners and meetings in Princeton and New York with Bogdon Maglich, the head of Migma Fusion, the only private nuclear fusion research operation in the United States, and a number of Yugoslavian government officials, I agreed to help organize a large Tesla celebration. To this end, I enlisted the support of the president of the prestigious Franklin Institute in Philadelphia, Bowen Dees, and after a stint at Harvard and with the blessing of the Yugoslavian Consul-General in New York, I went off to Yugoslavia, to spend days at their expense, as an unofficial ambassador.

I was planning to do many things during this celebratory conference that would have linked the Tesla Museum in Beograd with the Franklin Institute in Philadelphia: besides giving Tesla his just due and showcasing his achievements in a major exhibit at the Franklin Institute, while holding a major international conference on his works, I would also have organized a smaller conference on the suppressed aspects of his work in mind control and free energy and found a way to directly demonstrate mind control to those who came to the conference.

In the Fall of 1978; I was a Fellow in Residence at the Institute of Politics within the Kennedy School at Harvard. I taught one course, ran a small lunch time chat series in which Harvard luminaries, Ambassador Reischauer, E. D. Wilson and Karl Deutsch, among others, ate and chatted with 5 or 6 of us for a couple of hours; I lectured in every conceivable venue at Harvard, conducted a number of public symposiums, brought a number of the members of my Network to Harvard to lecture, ate dinner with a host of well known political figures, and made an inordinate amount of noise about mind control technology and the Russian Woodpecker to, among others, then CIA head Stansfield Turner.

This lead to a meeting in the Boston Airport, arranged by one of JFK's chief aides, on the matter with a top defense intelligence scientist who ended up spending the evening with me and giving me his home telephone number.

In 1979 I received a small private foundation grant to study free energy devices in preparation for a large involvement in such activities.

I was planning to visit all of the inventors personally and then prepare a report that would have formed the basis of a venture capital enterprise that had been encouraged, due to some of my mailings, by a number of my affluent friends. The objective was to develop and bring one device to the marketplace.[177]

In late winter of 1979 I went to Yugoslavia as a guest of the Yugoslavian government to further my Tesla Project. My translator was the same translator that Marshall Tito, the leader of the country, used. He spent much time with me outside of the official meetings, talking constantly about Yugoslavia's need to find a way to open to the United States and the West in general. I had a number of official meetings, the outcome of which was that the government agreed to fly whatever I needed of Tesla material from the Tesla Museum in

---

[177] Jack's comment: none of this would have worked I am quite sure. Ira was being used.

Beograd to the United States at their expense. I was also encouraged to return to Yugoslavia during the Summer and see more of the country at their expense.

Reports of my success reached both Bogdon Maglich and the president of Pennsylvania Bell, Bill Mowbraaten, before I got back to the United States. Maglich was amazed at my treatment. He couldn't quite believe the notes of the meetings he had received.

Bill Mowbraaten (sp?) had a message waiting for me when I returned to my small apartment about a breakfast meeting as soon as I returned. At breakfast the next morning, he told me that he wanted to take me to the next meeting of the Conference Board as American business was looking for a way into Yugoslavia and I had made a breakthrough. The Conference Board represented all the major American corporations and would have funded my conference and seen to it that the president delivered the keynote address. I was much too busy to think about how Bill had gotten information so quickly about my Yugoslav meetings, and I was immediately off to England again for a week with the Prince of Iran.

I was up to my ears in projects: I was in the middle of doing an interview with "Omni"; I was about to conduct a book length interview with Arthur Koestler that would have covered his entire career; I was in negotiation about acting in a play; the day my first life ended, a friend was coming down from New York to speak to me about helping to write and possibly act in a TV series about the 60's.

All was not to be. I was busted for a murder I did not commit and all my work on mind control and free energy became history."

[signed] Ira Einhorn, September 1, 2002[178]

From: <User886114@aol.com

To: <gear2000@lightspeed.net

Subject: Re: psychotronic "warfare" developments

Date: Saturday, June 30, 2001 8:37 AM

"David,

Psychotronics implies a mind component in the weapon system itself, not just the fact that the human mind is the target. ELF affects the human mind at very discrete and selective frequencies. That is what my original article in Co-Evolution Quarterly was about. That is what the Russian seemed to be doing with the Russian Woodpecker and we backed up our conceptual analysis that included a wide-ranging group, by building a small model and testing it on ourselves and in restaurants. It worked. The intelligence agencies of a number of western countries were informed, but were freaked by the entire idea or playing doggo. I spoke to the then director of the CIA about it while I was at Harvard and spent hours at Logan airport with a top intelligence scientist on the matter soon after that.

My trip to Yugoslavia was partly as a result of wanting to air the matter, by building a larger transmitter and demonstrating the technology to and ON top people, including the president who would have opened the conference. According to sources the CIA admitted in 1985 that the Woodpecker had mind control effect.

Another member of my network employed Adey, Persinger and a number of others, some of them part of my network to build a working device; when he tested it to his

---

[178] It was alleged in an AP report of Oct 8, 2002 on the Einhorn Murder trial that Ira had sought a book on how to actually make mummies about six months prior to his arrest. Of course, this does not make much sense since Holly's body would have had to mummify by then in the heat of the trunk in the storage room outside Ira's apartment. Ira denied the allegation. **Ira was convicted of the murder of Holl Maddux and given life without parole. Based on the new evidence presented at the trial, I personally suspect that justice was done.** (added 10/21/02)

satisfaction, upon himself, they closed down his lab and another lab doing similar work and he eventually lost his job. This was in the 80s. In the 90s, he and a past director of the CIA were consultants on a book that Larry Collins wrote about Soviet Mind Control, and published as a novel. So given 15 years in a black field, it is not hard to imagine working weapons. Mind control seems to upset the public more than anything else, BUT it also produces claims in purported victims that are over the top. The technology is real. I am sure it has been tested. I have no idea how effective it is, but if you can modulate brain waves, you can cause effect.

My article stirred up an enormous can of worms because of the place in which it was published and I was circulating information about such things to a world elite in 26 countries on both sides of the IRON CURTAIN in the middle of the COLD WAR.

The interest in things psychic re: weaponry was immense from 73-79 when my working life ended. The West was handicapped by its fear of looking foolish and terrified that they might be caught out in an area they had explored in other ways in the 50 and 60s. LSD, Sidney Gottlieb, the Olson affair (guy who jumped out a window).

There was intense interest among some of my Bell people who fed me stuff that I now think was a way of trying to elicit response. Remember Bell ran Sandia labs and at that time was unequaled as a research lab: transistors, lasers, et.al. I had access to everyone in the Bell system, brought URI[179] to Bell labs and was eventually supported directly by the NO 2 man in the entire organization: The Vice Chairman.

They were amazed at my range of contacts and happy to support my network for the privilege of sitting in on it.

No money changed hands and hundreds of key people got top flight key info for nothing." — Ira Einhorn

My comment on the above: Ira's claims are too vague to really evaluate. That there are effects of electromagnetic induction near fields and also sound waves on the human brain and behavior is a complex subject that requires another book. That effective mass mind control as a weapon beamed from a great distance is possible is very unlikely. There seems to be a lot of New Age hysterical paranoia about Harp and Woodpecker around this issue. However, the fact that naïve New Age people, who uncritically participate in street protests on practically anything, believe in Cargo Cult pseudo-science is indeed a potent psychotronic weapon more powerful than any beamed electromagnetic and sound waves. The use of ELF modulated microwave beam devices for clandestine bio-weaponry in small areas may be possible. I am not qualified in this specialty, but I do not automatically debunk it. The claims Ira makes here should not be confused with obviously idiotic bogus claims of zero point energy anti-gravity propulsion[180] other than what I have to soberly say about that in my field theory in this book. Most of what is written on "anti-gravity" and "free energy" in popular New Age books hogwash. None of it should be taken on face value.

Excerpts from Ira Einhorn's 1999 statement on his innocence

"FACTS FOR OPENERS

---

[179]Jack's comment: Uri Geller, e.g. "Magic and Paraphysics" by Martin Gardner.
[180] Most of these claims belong in the category of Baron Munchausen fantasy.

The most important aspect of any murder investigation are the forensics: the scientific reports connected to the physical evidence in the case.

If the forensics do not agree with the prosecutor's theory of the case, the case is almost certainly a loser for the prosecution.

The Assistant DA assigned to my case sent the physical evidence to the FBI instead of using the local expert whose job it was to handle such matters.

The FBI turned in extensive reports that refuted every contention about the case that the DA made.

These reports were withheld from us for a long period of time – though they were due to us by law.

STRIKE ONE

Then the evidence was sent to the prestigious scientific lab, National Medical Services of Willow Grove, PA, that did the work on the O.J. Simpson case, with the same result and the same delay in reporting to us.

STRIKE TWO

Then in desperation the evidence was turned over to the man whose competence was certainly questioned by being skipped over twice.

He turned up some protein by using a test that he invented and that would have been deeply challenged.

So not strike three, BUT this is not a baseball game, but a murder investigation in which there is normally one report, not the kind of shopping around described above.

When the negative results were finally given to us and announced, the city's major magazine replied with an article that had blood on every page.

There is no blood in the case. A badly smashed skull, which should spray blood everywhere, as my jubilant lawyer explained to me, produced no blood anywhere in the apartment. There was no blood in the body.

The conclusion of my lawyer was that Holly Maddux was not killed in my apartment. It was the off the record conclusion of one of the FBI men who did some of the forensic work. No wonder they went for a second opinion. It was also the conclusion of Kit Green[181] who at that time was Director of the Life Science's Division of the CIA who is an expert in this area of knowledge and who looked into the case.

The media went bat shit when I was arrested, came to the conclusion that I was guilty and then resolutely refused to deal with any evidence that contradicted their original assertions and conclusions, BUT did deal with the shopping habits of my wife at Friday market in Champagne - Mouton[182], so that the people of Philadelphia could read about my dinner the same day that I was eating it.

---

[181] Jack's comment: I have good direct reason to believe that Dr. Green does not agree 100% with Ira's assessment of his opinion and involvement in his case, which was very tenuous and cursory. There are false rumors about Dr. Green and Ira's case circulating on the Internet that caused Dr. Green anguish and possibly professional harm. I do not mean to imply here that Ira is deliberately falsifying or that he is guilty, though he might be, and he might also be innocent. My role and motive in all this is purely as fair witness in the interest of justice and fair play not as judge, jury and executioner. Since Ira was the literary agent for the original book in this series it would be morally wrong for me to ignore his story.

[182] Jack's comment: The region of my alleged ancestor Solomon ha-Zarfati a thousand years ago at the origin of the Legend of The Grail – French-German version.

What the prosecutor, Barbara Christie did however was a lot worse. She had a report in her possession that placed Holly Maddux in a bank and identified by three witnesses, six months after her supposed death at my hands. To keep this discovery from us, she shortened the pages of a larger report that was given to us, enabling her to avoid the pagination in the report and pull the incriminating pages out of the report. My lawyer smelled a rat and we eventually got the pages, BUT only after having to file a motion on the matter.

This was news. Not a word about it in the press, so then and there I got the drift of things, lost all respect for the law as it was being practiced in Philadelphia and began to think about another life, which became a reality when a third factor emerged that decided the issue. A factor I can't talk about as it involves a very paranoid person.

Barbara Christie built a career on such fraudulent practice which is now coming back to haunt her, as this recent e-mail note from a lawyer friend indicates:

—— "Just yesterday and publicized today because yesterday was a holiday - Veteran's Day - came the PA Supreme Court decision (in Com. v. Raymond Martorano and Albert Daidone Nos. 0002 and No. 0003 E.D. Appeal Docket 1998) both of whom were convicted of the murder of union leader John McCollough (Roofer's) in 1984. The prosecutor was Barbara Christie and she carried out a pattern of prosecutorial misconduct which resulted in the first degree convictions being reversed and now, the PA Supreme Court has held that her conduct was so egregious and so calculated to deny a fair trial that double jeopardy under the state constitution bars a retrial. While the decision was 4-3 it is still a real blow to the DA's office and to Christie. Because of her conduct - and her conduct alone - two mobsters cannot even be tried. " ——-

The behavior of the major American media has been totally consistent in following a party line which has presented me as a demon whose past life has been almost totally erased. An erasure in keeping with Joel Rosen's summation to the jury in my IN ABSENTIA trial as "a bum who xeroxed things" and members of the Philadelphia Earth Week Committee who signed a petition declaring that my participation in Earth Day was limited to 15 minutes in the face of all the evidence of my having organized, structured and been the Master of Ceremonies for the massive outdoor event which drew 250,000 people to Fairmount Park, lasted 7 hours, put ecology on the map and helped create the EPA.

... deleted sections...

DISCLAIMER: I am innocent of the charges lodged against me, BUT the above is not about guilt or innocence, it is about how I have been treated in a supposedly free press and the failure to deal with issues in article after article that lays shame on American journalism. A shame that so called Social Justice writer Russ Baker must also bear, for his ludicrous attempt, in his article on me in the December Esquire[183], to assign cause to some remarks on CIA involvement that were never offered as explanations, since I never offered Russ any explanations, as that was not the stated reason for his article. A frustration that ended in his lying about his failure to get any statements on the matter from me. I still await an honest journalist who will treat the matters outlined above in an adequate way.

Ira Einhorn

---

[183] Jack's comment: My name is in that Esquire article of December 1999 with a nude photo of Ira. Joe Firmage is alluded to as the potential provider of money to Ira to work on the French Military UFO report translation. Joe and Ira were in touch by email as part of ISSO operations in 1999 – 2000.

Jack Sarfatti
Champagne-Mouton, France
November 27, 1999"

"Dead Book" Author Hank Harrison (Courtney Love's Father) on Ira Einhorn

http://www.herzer.org/gallery/20020317HankHarrison
http://www.suitelorraine.com/suitelorraine/Pages/kingdavid.html
http://www.canoe.ca/JamMusicArtistsN/nirvana_cobain.html
http://www.heavenlyrain.com/cr/woman/faq/clovefaq.html

Hank Harrison, a Caffe Trieste Irregular at the Table Round and occasional houseguest wrote to me on 9/30/01:

"Jack: I have been silent on the Ira issue, but here is a brief response after reading the Salon.Com article.

'And Einhorn, who has always maintained his innocence, may take the witness stand to defend himself.'

Using the word "may" here is absurd. One could bet the barn on Ira's self-aggrandizing loquacious braggadocio. Ira would never miss a chance to rap and rant in public, especially if the focus is on him, even if it means a long sentence in prison. Several of his quotes on futurism, made no sense then and make none now. The height of his psychotic primadonna-hood came when he named his first book after its ISBN number. While the rest of us were virtually begging to be heard, especially here on the West coast, Ira was enjoying the praise and bafflegab of the East coast demimonde ... praise, he did not deserve, as none of his ideas were original, twisted; oblique; dense and opaque; but not original.

I have evidence that Ira snitched on people while in exile, and I did not like him even when I knew him as Ben Moore. He completely fooled me. Remember I had no knowledge of the Holly affair when I was in Ireland and did not know Ben Moore and Ira were the same guy until 1988 when Steve Levy met with me at Mac World expo. Moreover I copyedited Levy's second edition and my changes went into the paperback. I didn't like Levy either, but at least he was an excellent journalist with hundreds of articles in print and he worked everyday. I first met him when we were both on staff at InfoWorld. Levy was forthcoming and offered me a modicum of respect for my underground skills and associations. He was disappointed because he didn't win a Pulitzer, but that's common with east coast writers. Moreover, he seemed to be in competition with his live-in girl friend who, did, win a Pulitzer. But that does not distract from the veracity of his book, which I find, from my own research to be, mostly true. Levy does not reveal that he inherited a preliminary manuscript from someone who really knew Ira. Levy, to this day refuses to entertain any idea of Ira's CIA spook affiliations, but you and I both know they existed before and after Holly. Be that as it may, Levy did an OK job.

As to the body in the trunk being anyone other than Holly, forget it. There was no DNA test then, but finger prints nail the coffin. The idea of the smell of decomposition lasting a short while is also absurd.[184] The decomposing smell can easily be reactivated when dried fluids come into contact with a leaking water pipe or a drip from a rain soaked roof. As to

---

[184] Jack note: I independently confirmed this with a noted forensic scientist who disputes Eldon Byrd's public defense of Ira Einhorn on this particular matter.

the lack of blood on the bottom of the trunk, there was none, it went down the drain and was bleached out.

DeBenedetto, the Philly DA tracker, sent me copies of most of the forensic documents including pictures of the reconstructed skull. I still have them. DeBenedetto did most of the work and was then offered a Golden Parachute shortly after Ira was apprehended."

hh

In a message dated 09/29/2002 12:57:13 PM, sarfatti@well.com writes:

<< Ira was a complete slob and could not have cleaned his own apartment, >>

"More marginal assumptions. He would clean if he was on adrenaline...and he had a cleaning lady who could come in the next day. He had a year to clean it. The body was found more than a year after she mysteriously disappeared from the corner store. Remember also that she was lured back by Ira, that she left him months earlier, for what? Abusive conduct and philandering ... she met another guy in New York. Worked at an alternative Bakery, and only came back to Philly to get a few items still stashed at Ira's place. Plus he begged her to come back. Ira didn't like being dumped...who does? So much for Philly. There's another defense... A pure unadulterated testosterone attack.

In Dublin the place was always neat as a pin. Even when he did not have a lady living in." — Hank Harrison, 9/30/02

———-Original Message———-
From: Hankskids2@aol.com [mailto:Hankskids2@aol.com]
Sent: Wednesday, October 02, 2002 11:40 AM
To: sarfatti@well.com
Subject: Re: Was Ira a serial killer?

"In a message dated 10/02/2002 8:22:52 AM, sarfatti@well.com writes:
'Ira could not have murdered Marcia Moore[185] in I think it was California.'

Wrong again ... he was here for a visit to Bolinas and Big Sur on several occasions, but I do not have exact dates. Jack may have dates for his later visit to SF; I know he was in Big Sur, about 1973 and again later.

Jack: 'I think Marcia was killed, also by blows to the head (?) in 1979?'

Hank: In addition Steve Levy documents a dead woman's body found in the basement of his old apartment house in Philly, the Apartment before the infamous trunk apartment.

Jack: 'This allegation I did not know about.'

Hank: Moreover, folks, the padlock and hasp on the porch door after Holly disappeared were new, had only Ira's fingerprints on it, and was shiny bright. How could he have not seen it when he washed his cloths? The man claims to be hyper-observant. He claims he never went out there, he says he had his clothes washed at a full serve laundromat. But how could the phantom murderer know Ira never went there ... besides, the statement is not true because even Ira took his rubbish out every so often, he would have seen the new lock eventually and wondered about it — unless of course he put it there.

Dead bodies seem to crop up around Ira, how do we know Ira didn't torch Puharich's "Sing-Sing" house. There seems to be a bi-coastal connection here. Wasn't Puharich, at least for a time, living in Carmel? California.

Jack concluded by stating:

---

[185] Both Holly Maddux and Marcia Moore were members of Andrija Puharich's group in Ossining near Sing Sing.

'I don't know any more except that the person who killed her may have also killed Holly Maddux - that's my speculation as amateur sleuth.'

Then again you may be right. If Ira killed Holly he may have done some other mischief... the MO seems to be the same, why would a serial killer stop at arson? What was in the house that needed to be destroyed? ..."

Investigative Reporter Dick Farley wrote to me about Ira Einhorn on October 1, 2002

(excerpts follow):

"'Fun's over,' and with Ira on trial in a hostile media environment and totally corrupt legal context there are steps which I believe are possible to assist Ira in establishing a credible context for his beliefs in a conspiracy to discredit him. IF the judge will allow it, which requires credible witness testimony that Ira's attorney OUGHT to be actively recruiting from YOU and others in the previous Einhorn sphere of influence, a case for reasonable doubt can be made.

There is ample journalistic stuff "out there" which could tie things together. And there are "real people" on the record about the things Ira was involved in, at that time so "deep" and "weird" that they'd not have been believed but which, if they could be pushed through the media "screen" of disinformation agents and young nitwit folk not even out of puberty when all this went down (but who now are "gatekeepers" at the wire services and other news outlets!), might gain "legs" and substantiate Ira's view....

Question: Who is paying for Ira's defense? William Cannon, his attorney, seems to be playing the role of "fair trial" but certainly not aggressively. In cases like this, it falls to the accused "friends" and defense team to PROTECT the "family jewels." I would be most suspicious of my "friends" if I were Ira; they have the most to lose if the Truth (big T) were to come out. That's why they hid him. That's why we wanted him on the stand. What he says could be easily validated sufficiently to provide him with the "reasonable doubt" that the hateful Maddux family and the Philly DA want so desperately not to allow. Lynching Einhorn does not serve justice and it will not be helpful in bringing into sunlight and justice those who operated in the Dark Corners. ...

Whoever killed Holly and whatever Ira believes or recalls about that whole episode, it is important for Einhorn and his defenders to open up the coffin and expose to the sunlight those "vampires" who lurked along the fringes of the "New Age" and politics of the turbulent Sixties and Seventies....

The place to look, then and now, for what Ira was messing with is probably not only CIA, although their operatives were certainly all over his scene, but also at DIA and at USAF, particularly when it comes to the Tesla aspects of Ira's work. Also, in the book by Dr. Jim Schnabel, "Remote Viewers: The true story of America's Psychic Spying," or some such, published in about 1994 or 1995, there is ample narrative and quotes from Major Ed Dames (of Psi-Tech, chaired at one point by retired General Stubblebine and others still very present in "The Weird") about how DIA, under one Dale Graff (formerly USAF Intelligence at Wright-Patterson AFB., until he ran afoul of brass and was transferred to DIA at Bolling AFB, according to Schnabel and Graff's friend Jones, who "advised" Graff on what to do after his retirement in the mid-1990s from DIA, so said Jones), was allegedly involved in the very kind of "psychic spying" that Ira is claiming, to much media derision, was going on. Major Dames, who was one of the Remote Viewers in the NSA-cum-CIA "coordinate remote viewing" program and was featured in Schnabel's four-hour British documentary about all this which ran on an American cable network in the mid-1990s, is quoted by Schnabel saying that Graff and DIA, in the later stages of the "RV" and "applied anomalous phenomena" program, even resorted to "the Witches," as Dames said the military psychics at one point came to call the New Agers allegedly recruited by Graff, et al. Schnabel tells of an alleged romantic link between one woman "fortune teller" working for Graff, and a senior Senate Appropriations Committee staffer, Richard D'Amato. D'Amato worked for Senator Robert Byrd[186] and one of D'Amato's duties was oversight of some of the "black projects." See Schnabel's book for context that verifies Einhorn's "milieu."...

Item: Dr. Hal Puthoff published a monograph outlining his work at SRI in the 1970s on "psychic spying" techniques in the Journal of Scientific Exploration[187] of the Society for Scientific Exploration (SSE) a few years back. In it, Puthoff included a photo of three men standing in front of a small plane. Besides Dr. Puthoff, pictured were the psychic Pat Price and Dr. Christopher "Kit" Green, former senior CIA scientist who served as the Agency's contract manager for the work Puthoff, Targ, et al., did at SRI, with management at that end by the late Dr. Willis Harman. Now, Harman's book, "Global Mind Change," as it was published in 1977 in the UK, is instructive about the world view in which Ira Einhorn was dabbling. In Stephen Levy's book, a passing reference is made to Einhorn's visit to San Francisco and Palo Alto, where Levy says Einhorn met with Dr. Jack Sarfatti and with Dr. Jacques Vallee, both of whom were actively at work on what eventually became the Internet. Einhorn was, characteristically, very excited about the "global neural net" the Internet would be, as Ira considered himself a "planetary enzyme," allegedly working on behalf of what he at the time may have believed were "off-planet intelligences" of the sort described by Puharich and his protege, Israeli psychic Uri Geller, with whom Ira had contacts, as Levy again documents. What is ironic about Puthoff's monograph is that it "outed" the CIA's manager, Kit Green, whether inadvertently or not. The same photograph that Puthoff used in his SSE paper ran in Jim Schnabel's book, "Remote Viewers." But Schnabel had identified Kit Green by his cover identity, "Richard Kennett," as well as identifying Puthoff and Pat Price.[188] ...

The Puharich and "Council of Nine" link to rich New Agers is the part of Ira's story that NOBODY is interested in having come out. But it is the most interesting part, for which there is ample documentation in print (although obscure by the media today), and which threads a trail across thirty or more years into the salons of the "elite." And the Puharich thread ties many high-profile people with New Age links into this, from Shirley MacLaine to current Rep. Dennis Kucinich (D-OH) and "active others."[189]

Most current "players" in the fringe fields of psychic "remote viewing" and "ET/UFO" cadres are retired intelligence and military operatives, some working for rich patrons having a most disturbing (perhaps "disturbed," but I'm not qualified to diagnose ;-) world view about Earth and its future. It was here Einhorn found himself and made a lot of noise which

---

[186] Jack comment: Senator Byrd is currently arranging the Federal funding for Institute for Software Research in West Virginia with Dr. James Corum who, in addition to being an expert on Tesla's engineering, is also working on exotic propulsion physics like I am and for much the same reasons.

[187] Jack comment: Edited by Bernie Haisch at the time as I recall.

[188] Jack comment: Pat Price was a close friend of the late Harold Chipman who had been a CIA Chief of Station in Munich I think? I knew "Chip" in San Francisco in the mid 1980's.

[189] Jack comment: Like Joe Firmage who I put in e-mail contact with Ira back in France in 1999.

the elites, who at the time controlled some intelligence agencies, did not appreciate. They are not "The Government!" They work "beyond the fringe."

According to Schnabel, these people and their worldview troubled journalists of the day like Leon Jaroff, science editor at Time Magazine, who wrote to discredit them and the work at SRI, Schnabel reports. Schnabel footnotes in his book that some of the folks like Jaroff saw similarities in the "New Age" underpinnings of what the SRI folks were dabbling in with the underlying "New Age" pseudoscience that held so much fascination with the Nazis like Himmler and major Teutonic backers of Hitler. ...

Remember this: Justice doesn't only apply to those personalities we admire or like. If Einhorn is railroaded, or if the media "lynchings" under way by CNN and MSNBC do not provoke a challenge, then whomever "They" are protecting will only grow stronger and we, as freedom-loving people, will be further estranged from our civil liberties. It is irrelevant how I as a journalist think about Einhorn or my own "feelings" about what or who he may be. There are a lot of "you" out there, on this list, who know a LOT but who, for whatever reasons, are allowing Einhorn to 'hang alone.' ..."

Dick Farley <cloudrider@aol.com>
Washington, DC USA

I note Wes Thomas's Einhorn Archive http://www.sonic.net/~west/ira/
Yes, if they have such a strong defense they would be nuts to put Ira on the stand. Ditto for me. As soon as we started telling the real Truth about the paranormal and UFOs et-al they would execute Ira and put me in a padded cell with John Nash. Nope, I am not going near Philly where they have reporters like this Wizard Hunter Ms Conroy.[190]

Keep Ira from shooting himself in the foot. They have a strong case if George Draper does not get murdered before he testifies. I would not wish to be Draper right now. I hope he is packing a pistol and is a good shot. He needs to watch how he cross streets.

Seriously, unless it's a lynching, all Ira's defense need do is have former Philadelphia Police Officer George Draper say he saw Holly alive AFTER Ira allegedly murdered her. Case over. Ira wins and is back in Champagne-Mouton where my alleged ancestor Solomon ha-Zarfati AKA Rashi des Troyes (1040 - 1105) came from.

Ira is his own worst enemy right now. The simple folk on The Jury will not understand anything he says. So unless they want to cop an insanity pleas keep Ira and his equally nutty friends like me off the stand. With friends like me, Ira does not need enemies. ☺

On the other hand the circumstantial evidence against Ira seems damning, e.g.
"By John J. Goldman, TIMES STAFF WRITER Oct 2, 2002
PHILADELPHIA — Ira Einhorn jealously murdered his former girlfriend in a classic case of domestic violence, and even wrote a poem after he struck another ex-lover with a bottle and tried to strangle her, a prosecutor charged Monday.

The poem was titled "An Act of Violence," Assistant Dist. Atty. Joel Rosen told the jury during opening arguments in the trial of the onetime counterculture figure.

"Suddenly it happens. Bottle in hand I strike away at the head. In such violence, there may be freedom," Rosen quoted Einhorn as writing.

---

[190] Conroy's calling Einhorn a "professional freeloader" is not Pulitzer level.

'He had his own little bizarre philosophy of violence," Rosen said as Einhorn stared intently at the jurors. "This defendant believed in the use of violence when breaking up a relationship.'"

PSI WARS!

——-Original Message——-
From: Howard Fertman [mailto:hjf35@earthlink.net]
Sent: Wednesday, October 02, 2002 4:53 PM
To: Jack Sarfatti
Subject: RE: With friends like Crockett Einhorn does not need enemies.

Yo Yippy Dippy David, greetings from Philadelphia. So now we see the Einhorn trial cast as relevant to the Iraq conflict, regarding the so-called suppression of zero-point free energy technologies from recovered UFO's. My god, Ira could have saved the world if only he wasn't framed. He was too preoccupied with matters of the highest importance to have noticed the foul odor... he must have assumed it was his own foul body odor. Ya that's the ticket...over the 18 months he was mostly away and during the short time spent in West Phila, he just assumed that the foul odor of a rotting corpse was his own B.O.

I think the defense should call in Steven Greer and the whole [UFO] Disclosure cast. A mistrial should be called since the jurors have most likely been victim to psychotronic weapons, clouding their judgment. Ira stands between the truth and the abyss and the fate of the world weighs in the balance.

Jack, What happened to your 50 / 50 position? In light of the "web of intrigue"

that your circle basks in, how ridiculous to tie Frank Rizzo to this – to me, complete intellectual dishonesty.

Jack replies: Are you talking to me? http://www.filmsite.org/taxi.html You can perhaps fairly accuse me of gullibility and plain stupidity in my bending over backwards to come up with some sort of explanation after I saw "The Thin Blue Lie" http://www.moviereporter.com/reviews/thinbluelie.html , but of "complete intellectual dishonesty"? No way Jose

Howard: If there were an interest in silencing Einhorn, he would be dead. No sober mind would buy into a setup... you all have read too many comic books. It is so much more likely that Ira went off the deep end in a jealous rage. He is a controlling narcissist with a history of battering women, sucked up on delusions of self importance and addicted to intrigue, and probably was doped up to boot during that period when the murder occurred. Now what happened to Don Wiley, the Harvard Molecular Biologist, who's car was found empty and running, on the Hernando DeSoto Bridge, over the Mississippi River... in the weeks after the anthrax attacks....now that has the signature of real-world intrigue.

Yippy Dippy David,

If you're going to quote the Sarfatti emails, include all the positions – its clear your character is greatly lacking, your thinking is unbalanced and, by the way, keep up the good work, I'm sure the prosecution is very pleased with your efforts.

—- Howard Fertman
—- hjf35@earthlink.net
—- EarthLink: The #1 provider of the Real Internet.

Jack Sarfatti

Jack: That's what I also told David. The New Age is the Dumbing of America with "toxic cosmologies"[191], "Cargo Cult Pseudo Ccience"[192] and Sham Spiritualism by modern Elmer Gantry's.

——-Original Message———-

From: David Crockett Williams [mailto:gear2000@lightspeed.net]

Sent: Monday, September 30, 2002 8:53 PM

To: IRAlist

Subject: Ira's lawyers hope to keep his defense real

http://www.philly.com/mld/dailynews/4180399.htm

Posted on Mon, Sep. 30, 2002

Ira's lawyers hope to keep his defense real

That may be difficult if he testifies

By Theresa Conroy

conroyt@phillynews.com

"For nearly 25 years, Ira Einhorn has offered an outlandish defense for how the body of Holly Maddux wound up mummified inside a steamer trunk in his apartment.

You see, it really was the CIA, FBI - maybe even the KGB - that killed Maddux and then put her body in the trunk, just to discredit Einhorn's groundbreaking work on psychic mind control, Einhorn has said.

To add even more intrigue through the years, Einhorn padded the story with tales of UFOs, quantum physics, psychic phenomena, spoon-bender Uri Geller and telepathic transmissions from Russia.

But with the "planetary enzyme's" murder retrial set to begin today, it's time for the rubber to meet the road - or, at least for the landing gear to meet the surface of the moon.

William Cannon and Mitchell Strutin, the court-appointed attorneys defending the professional freeloader, want to keep their courtroom strategy as close to Earth as possible.

'We're not going to try to prove who killed Holly. That's not our job. That's their job," Cannon said. "They have a theory of the case which we intend to attack.'

"Their theory is that on or about Sept. 11, 1977, Ira beat her to death in his apartment and put her in a trunk and there she lingered until her body was discovered 18 months later," he said.

'Everything that makes that difficult to swallow helps us...We have an array of things to contradict that.'

Cannon said he intends to instill reasonable doubt in the jury by questioning three witnesses who say they saw Maddux months after Einhorn allegedly beat her to death.

Two of those witnesses - a bank teller and bank security guard – testified during Einhorn's 1993 absentia trial but waffled under cross-examination. The new witness, former Philadelphia Police Officer George Draper, will make a strong statement, Cannon said.

'He's adamant that he saw her alive and well in March of '78,' Cannon said.

'This guy's unshakeable.'

Cannon also plans to poke holes in prosecutor Joel Rosen's case by presenting expert forensic testimony that Maddux was not killed inside Einhorn's Powelton Village apartment.

The scenario could work, except for that wild card: Putting the fanciful guru on the stand.

---

[191] Coined by Dick Farley.
[192] Coined by Richard Feynman.

Cannon said Einhorn is 'prepared' to testify if needed.[193]

Other prominent Philadelphia defense attorneys said they doubt Einhorn's attorneys will be able to prevent him from taking the stand.

'You can suggest to a client that it may be your opinion not to take the stand; however, that decision is left solely up to the client," said attorney C.P. Mirarchi. "You can stand on your head and say, no, you're not taking the stand, you're not taking the stand. He has the final say. I think it's going to be hard for [Einhorn] to say, no.'

That's when things get dangerous, Mirarchi said.

'No matter how you rehearse it, if during the cross-examination by Joel [Rosen], he starts going off on a tangent because Joel knows where the bottom line is and starts asking it - starts asking about the Tooth Fairy and little green men and the CIA - he's going to feed into it.'

If Einhorn does get into mind-bending territory, 'then he's on his own to explain that,' Cannon said.

He hasn't had much success with that in previous court appearances.

During one of Einhorn's French extradition hearings, the fugitive took the witness stand only to spend 15 minutes rambling about 'Star Trek,'[194] his friendships with pop singer Peter Gabriel and Yugoslavian officials and how scientists all over the world were counting on his research to save the planet.[195]

Looney or not, Einhorn has to testify in his defense, said criminal defense attorney Todd Henry. But, Henry said, those hours - or, in this case, perhaps, days - on the stand don't have to harm Einhorn's case.

'I think in this case, realistically, he has to testify because he's an enigma, from what everybody has read about him,' Henry said. 'Many of [the jurors] may not even remember this case. They don't know who Ira Einhorn is. They don't know what he was and what he meant to the city, to the country, and - in his crazy mind - the world.

That can be used well in your defense. The guy gets up there and comes off totally wacky but can give you some tidbits that you can hang your hat on.'

To offer credible tidbits to the jury will require that Einhorn remain focused - something the perpetual blowhard may not be able to do.

'If Mr. Einhorn testifies, there is a danger he will forget that he is on trial for his life and view the cross examination as a match of wits,' said attorney Jules Epstein. 'And that could prove devastating.'"

Received Oct 2, 2002 from Davey Crockett:

"THE INVISIBLE THIRD WORLD WAR"
by W.H. Bowart and Richard Sutton 27.9.1990
http://www.angelfire.com/or/mctrl/bowart.htm#72

---

[193] Jack comment: Big mistake in my opinion. Ira is deluded on all the phony physics stuff he got from Tom Bearden and others. Ira's only relevant defense is George Draper. The circumstantial evidence against Ira is very strong.

[194] Jack comment: Ira was probably correctly alluding to fact of Andrija Puharich's shaping of Gene Roddenberry's ideas for Star Trek. Roddenberry got spooked by all the high weirdness around Puharich and split before I met Puharich and his gang. Curiously, if my physics is correct it will "Make Star Trek Real". I do not know yet if it is correct, but there is a plausible chance that it is.

[195] Jack comment: Ira is deluded about that like John Nash was. The fact that Ira is crazy does not make him ipso-facto a murderer.

"XIV. MAGICAL MYSTERY MURDER

Throughout the globe, an underground physics network acts as a thorn in the side of the invisible warriors, studying the latest developments of paraphysics and openly publishing its findings (60). Such colorful characters as Bob Beck of Los Angeles' Biomedical Research Associates fly back and forth from Washington to Eugene to Canada, sharing findings and investigating suspected government environmental crimes. Members of this loosely-knit network share their information freely and are concerned about the mind control aspects of the invisible war.

A former military intelligence officer, Lt. Col. Thomas E. Bearden, USAF (Ret.), publishes SPECULA, a magazine dedicated to "psychotronics" and "bio-energetics". These two words describe what amounts to the electronic amplification of telepathy and what has heretofore been called ESP. Bearden is not just a concerned amateur, he is an experienced scientist who has several of his psychotronic weapons papers on file at the Defense Documentation Center outside Washington (61). The story of his fight to publish a book on the subject of psychotronics and bioenergetics, THE EXCALIBUR STATEMENT, is as full of cloak-and-dagger intrigue as an Ian Fleming[196] thriller (62). Bearden's book was due out in 1978, but has been "mysteriously" delayed. During the time he was preparing the manuscript for publication, one by one, the members of the "network" met strange fates.

One such incident is the Ira Einhorn Murder Mystery. Einhorn, an organizer of the "network", directed some of Congressman Rose's[197] efforts from behind the scenes at the House of Representatives. One day Einhorn was shocked to discover the long dead body of his fellow researcher in a trunk on his back porch (63). When murder charges were pressed against Einhorn by the FBI, Einhorn launched a passionate defense which resulted in a great press controversy and a long court battle to get the evidence against .... (text missing) and currently Einhorn is hiding, convinced he was framed. But who framed him? It could have been U.S., Soviet, or British Intelligence (64) since Einhorn freely exchanged information with any group or individual who requested it. In hindsight, he suspected that some people who requested hard to get Tesla papers may have been working for foreign intelligence agencies. We examine the FBI case documents and other investigations which pose some intriguing questions about the Einhorn case."

My comment for the record: I have never been able to understand Bearden's "physics" and I do not believe his claims on "free energy" devices. Ira Einhorn was not a physicist and could not evaluate the accuracy of Bearden's claims. However, disinformation and misinformation on these subjects was a tool of covert Psy Ops on both sides of the Cold War. Dennis Bardens[198] at the Blue Boar Inn in Cambridge England in the spring of 1974 told me:

———-Original Message———-
From: Wes Thomas [mailto:west@sonic.net]

---

[196] Jack comment: My late friend Marshall Naify was one the financiers of Fleming's James Bond films with Sean Connery. Indeed, Marshall had one of the four black Aston Martins from the films.

[197] Jack comment: Charlie Rose, DNC House Select Committee on Intelligence who also was part of the Esalen Hot Tub Diplomacy with G. Arbatov's Institute of US and Canada.

[198] Bardens was later identified to me by the late Peter Maddock http://www.imprint-academic.demon.co.uk/SPECIAL/07_01.html as "a part-time stringer for British Intelligence". Maddock was at the paranormal meeting of Ted Bastin's group that I attended with Brendan O Regan, Chris Bird and Joyce Petchek. I met Brian Josephson at this meeting. Bardens wrote "Ghosts and Hauntings" and "The Ladykiller: The Life of Landru, the French Bluebeard". This last fact is curious in the light of the Einhorn mystery.

Sent: Friday, October 04, 2002 4:07 AM
To: RTSData@aol.com
Cc: Jack Sarfatti; User886114@aol.com
Subject: WARNING TO IRA'S DEFENSE TEAM!

Rich,

I think physicist Jack Sarfatti (a leading expert on esoteric physics) makes a valid point in the message below. Ira's conspiracy ideas, especially on mind control (which I'm an expert on, by the way; I run an Internet email list on this subject) and free energy will be seen as ridiculous and unconvincing. It's not a realistic defense and he's going to come across as a dangerously deluded wacko. This could seriously harm his credibility and destroy your case. You need to get to Cannon on this. He needs to talk to Jack or other top physicists and get an instant education on this stuff.

Jack: Especially the "free energy" and also idiotic claims of beaming viruses on microwaves like Laputan "sunbeams from cucumbers". The Einhorn Case, like the Rise and Fall of Joe Firmage (especially The Fall), show the real dangers of deluded New Age thinking in which Cargo Cult pseudo-physics, sham self-help spirituality and simplistic kitsch knee-jerk imbecile leftism replace critical thinking, excellence and real knowledge.

Subject: [ira-einhorn] RE: Ex-neighbor tells of Maddux crying on steps

Jack wrote: For the record: I agree 100% with Jafolla:

"Jafolla, a physicist now living in San Diego, said that in the mid-1970s Einhorn tried to engage him in discussions about "unusual" physics and espionage uses of microwaved viruses… It was nonsense physics,' Jafolla testified dismissively, as Einhorn, 62, raised an eyebrow." By Jacqueline Soteropoulos Inquirer Staff Writer Oct 3, 2002

Indeed, that's Lt. Colonel (ret) Tom Bearden's stuff. Some kind of Cold War disinformation I suppose with Ira as the patsy perfect "useful idiot" who does not know when he does not know. It looks bad for Ira with that defense attorney not sticking to a simple defense with the retired cop Lt. George Draper who says he saw Holly alive after the alleged murder.

"Dr. Sarfatti, first I want to tell you that I am a Cabalist. It is my duty to inform you of a psychic war raging across the continents between the Soviet Union and your country, and you are to be in the thick of it."[199] Dennis Bardens, Blue Boar Inn, Cambridge, UK, Spring 1974

---

[199] My book, "Destiny Matrix" (1st Books, 2002) has many more details, documents and photos on this topic of covert intelligence operations of "Hot Tub Diplomacy" during the Cold War using the New Age Human Potential Movement centered in the Bay Area and Esalen in Big Sur as well as in Britain, France and the Soviet Union. Michael Murphy, owner of Esalen, and one-time close friend of Ira Einhorn, also wrote a "novel" about this called "An End To Ordinary History" that Saul-Paul Sirag helped him with on the physics. See also "The New Mental Battlefield" by Colonel John Alexander in Military Review, 1980. About ten years later, British Psi Researcher Peter Maddox, who was at the 1974 Cambridge Psi Meeting, told me that Bardens was "a part-time stringer for British Intelligence".

Index of key words

ABOUT THE AUTHOR

SAN FRANCISCO, CALIFORNIA—"The Bohemian physicist…contributes a balanced scientific non-establishment for this expanding society. I don't mean to disparage the work, either…among all the blatherers there sometimes appears a breakthrough thinker. Originality has always required a fertile expanse of fumble and mistake. That's the beauty of the option. Your life might turn out to be just what's required to save the planet…Sarfatti's Cave is the name I'll give to the Café Trieste in San Francisco, where Jack Sarfatti, Ph.D. in physics, writes his poetry, evokes his mystical, miracle-working ancestors, and has conducted a several-decade-long seminar on the nature of reality and his own love life to a rapt succession of espresso scholars." *Bohemia, Where Art, Angst, Love and Strong Coffee Meet*, Herbert Gold, Simon & Schuster, 1993.

"Jack Sarfatti, Ph.D., Director of the Physics/Consciousness Research Group, is the catalyst without whom the following people and I would not have met … These men are the godfathers of this book." Gary Zukav in Acknowledgments to *The Dancing Wu Li Masters*. Jack, as a matter of fact, actually wrote large parts of the physics sections of Wu Li Masters and worked closely with Gary on the re-writing of contributions from other physicists whose cooperation Jack brokered. Gary had no physics education and was Jack's roommate in San Francisco's Bohemia – North Beach, which is how he got the opportunity to come to Esalen because of Jack's invitation that led to the Wu Li book.

Author Jack Sarfatti was born in Brooklyn, New York on September 14, 1939. He graduated with a BA in Physics from Cornell and later earned a PhD in Physics from the University of California. Jack has been President of the Internet Science Education Project in San Francisco since 1995 and has received substantial no-strings genius grants from several sources. In addition to his active research and prolific writing, he currently administers funds donated to leading cultural institutions for several scientific international research projects ranging from theoretical physics to nanotechnology and archaeology.